LAB MANUAL

to accompany

General, Organic, and Biological Chemistry

TODD DEAL
Georgia Southern University

General, Organic, and Biological

Chemistry

Frost

Deal

PEARSON

Boston Columbus Indianapolis New York San Francisco Upper Saddle River
Amsterdam Cape Town Dubai London Madrid Milan Munich Paris Montréal Toronto
Delhi Mexico City São Paulo Sydney Hong Kong Seoul Singapore Taipei Tokyo

Editor in Chief: Adam Jaworski
Executive Editor: Jeanne Zalesky
Senior Marketing Manager: Jonathan Cottrell
Project Editor: Jessica Moro
Editorial Assistant: Lisa Tarabokjia
Marketing Assistant: Nicola Houston
Managing Editor, Chemistry and Geosciences: Gina M. Cheselka
Project Manager, Production: Edward Thomas
Full-Service Project Management/Composition: PreMediaGlobal
Illustrations: PreMediaGlobal
Text Permissions Manager: Alison Bruckner
Text Permissions Researcher: Liz Kincaid, PreMediaGlobal
Cover Designer: Seventeenth Street Studios
Operations Specialist: Jeffrey Sargent
Cover Image Credit: Mark Conlin/V&W/imagequestmarine.com

1 2 3 4 5 6 7 8 9 10—VHO—16 15 14 13 12

www.pearsonhighered.com

ISBN-10: 0-321-81925-X
ISBN-13: 978-0-321-81925-3

Table of Contents

Preface

This laboratory manual was written to accompany *General, Organic, and Biological Chemistry, 2e*, by Frost and Deal, but is designed to accompany the laboratory portion of any one semester general, organic, and biological chemistry course. The majority of the experiments include a link to the health sciences, such as nursing and nutrition, while the concepts highlighted are framed in real-world questions and are broadly applicable. Users of this manual will discover that many of the experiments illustrate concepts from more than one chapter of the text and often utilize foundational knowledge from the areas of general, organic, or biological chemistry to develop concepts in one or more of the other areas. This integrated strategy, which is the cornerstone of the text and of our teaching method, helps students to understand that chemistry is not a disparate set of unrelated concepts. Using our integrated approach, students develop a seamless framework to help them understand chemistry and to see its applications in their everyday lives.

DESIGN OF EXPERIMENTS

Chemistry is an experimental science and as such is built upon inquiry. As chemists, we ask questions about phenomena we observe in the world and then we take those phenomena into the lab and investigate them. This lab manual is designed to engage students in the process by which we investigate the world and practice our craft—the process of inquiry. The majority of the experiments in this manual pose a question for students to consider and investigate. I have attempted to frame the questions in ways that will engage students and demonstrate the relevance of chemistry to their everyday lives.

Components of each experiment:

Each lab begins with one or more **Learning Goals** to guide students into the concepts they should be learning through the experiment. For some of the labs, I have reframed the learning goals and included them in the end-of-lab questions. As you assess students' learning in each lab, I hope you will use the learning goals to guide you.

The **Before You Begin** section of each lab is intended to guide students' pre-lab reading and to highlight the concepts they need to know before coming to the lab. For those students using the second edition of the text, the readings are tied directly to the sections of the text. For other students, I have included a list of concepts that they will need to read and understand from their text prior to the lab.

The **Introduction** section is intended to highlight the concepts that students will use and/or learn in the experiment. These sections are not intended to be an in-depth treatment of the relevant concepts and should be supplemented with appropriate readings from the text. After reading this section, students should understand the context of the day's experiment and have an idea of the concepts they will be investigating.

Each **Experiment** section provides step-by-step instructions that are tied directly to the **Report Sheet**. Students will be guided to complete the appropriate sections of the report sheet as they move through the experimental procedure.

The **Pre-Lab Questions** are intended to check students on their understanding of concepts relevant to the day's experiment and to check their reading of the material included in the Before You Begin and Introduction sections of each lab. Students should complete the pre-lab questions prior to coming to the lab. Students can remove the pre-lab questions from the manual and hand them in prior to the lab without compromising the information they will need to complete the lab.

Finally, the **Questions** are designed to check students' learning and to tie their day's work to the initial question of the lab. After completing the experiment and its associated analysis of information, students should be able to give a plausible answer to the inquiry question that frames the lab.

TO THE STUDENT

Welcome to chemistry lab! Perhaps you are that student who sees the laboratory portion of your chemistry course as one more obstacle in your week and you just want to get it over with. I wrote this manual with you in mind. Each of the experiments included here is designed to help solidify your classroom experience with the concepts included, but I have framed the experiments around a question or situation to make the experience more engaging and interesting. Perhaps you are the student whose favorite part of chemistry is the lab. I also wrote this manual with you in mind. You will find that each of the experiments challenges you to apply what you know to discover new concepts and to answer questions relevant not only to your classroom experience, but also to your everyday life. My hope is that chemistry lab becomes an enjoyable part of your week where you apply what you have learned to become a creative and critical thinker and an excellent problem solver.

In any laboratory experience, safety must be your first concern. Accidents in a chemistry lab can be particularly dangerous, but they are also easily avoided. The first experiment in this manual is intended to help you understand all safety precautions, especially those relevant to your laboratory. You should complete this experiment before beginning any of the others included in this manual. Prior to each experiment, read the day's experiment and take note of any safety precautions. Listen carefully to your instructor's pre-lab safety instructions and follow all lab safety guidelines. Always dispose of your lab waste as directed by your instructor. If you are in doubt as to how to appropriately handle any chemical or waste product in the lab, seek your instructor's guidance.

ACKNOWLEDGMENTS

A project such as this is never a one-person show, but involves the support and contributions of many people. I would first like to acknowledge the support of my family, who encouraged me to take on the preparation of this lab manual and gave me permission to steal away and write when necessary. This project would never have been born without the persuasive encouragement of my executive editor, Jeanne Zalesky. Jeanne's initial vision for this project inspired me, and her constant encouragement helped make it a reality. I am grateful as well to my editor, Jessica Moro. Jessica is a master of motivation who kept me on task and focused on the project. I still marvel that she was able to keep me moving and to finish this project ahead of schedule. I will forever be indebted to Jim Smith, who was the first editor to believe in the integrated strategy. Without Jim's encouragement, this lab manual and the accompanying text would never have come to be. Finally, I would like to say thank-you to the reviewers, accuracy checkers, artists, and compositors who worked behind the scenes to make this project a reality.

VISUAL GUIDE TO LABORATORY EQUIPMENT

Below is an example of most of the equipment you will use in a your chemistry laboratory. Familiarize yourself with this equipment and its uses in preparation for each laboratory period. The equipment as shown is a generic representation of typical lab equipment. Your equipment may look slightly different than what is shown here.

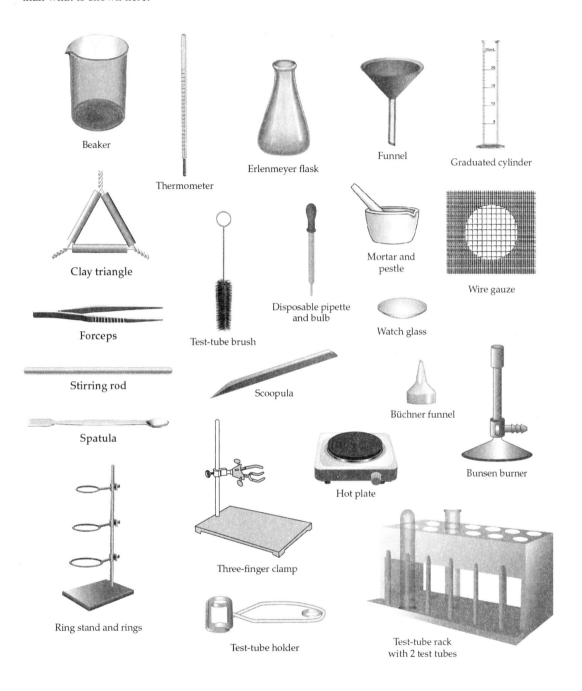

Beaker

Thermometer

Erlenmeyer flask

Funnel

Graduated cylinder

Clay triangle

Test-tube brush

Disposable pipette and bulb

Mortar and pestle

Wire gauze

Forceps

Watch glass

Stirring rod

Scoopula

Büchner funnel

Spatula

Bunsen burner

Hot plate

Three-finger clamp

Ring stand and rings

Test-tube holder

Test-tube rack with 2 test tubes

Why Is Safety So Important in Chemistry Lab? | 1

LEARNING GOALS

After completing this laboratory exercise, you should be able to:

- Identify the safety equipment in your laboratory and demonstrate the proper use of each piece of equipment
- Discuss the importance of laboratory safety and rules for general laboratory safety
- List guidelines for appropriate clothing and safety apparel for a laboratory setting
- Explain safe handling and disposal of chemicals, including the use of chemical hazard symbols and material safety data sheets (MSDS)
- Describe emergency procedures for a laboratory accident or chemical spill

BEFORE YOU BEGIN

Read the introduction that follows; then work through the pre-lab exercise at the end of this experiment to determine whether you understand the concepts necessary to complete this lab.

INTRODUCTION

Welcome to Chemistry Lab! As you will discover throughout this semester, chemistry is not a collection of facts to be learned, but is a way of observing your world, collecting and testing data on those observations, and interpreting the results of those tests. In other words, chemistry is an experimental science. As chemists, we typically carry out our experiments in a laboratory.

The laboratory can be an exciting and fun place to work, but it also can be a dangerous place. In this experiment, you will be introduced to important safety procedures and equipment designed to help avoid accidents and protect you from personal injuries as you work in the lab. You will also become familiar with sources of information and guidelines to follow as you work in the lab that will help ensure your safety as well as the safety of those who are working in the lab with you.

Read and study the following sections and the Commitment to Laboratory Safety Pledge that follows them. Sign the Pledge and submit it to your instructor before you begin the activities for today's lab.

Working Safely in the Laboratory

Never work alone in a laboratory. An instructor should always be present, and you should work with a partner if possible.

Do only the work assigned by your instructor and follow the designated experimental procedures as written or explained.

Do not engage in horseplay or rough play while in the lab.

Wear approved eye protection at all times. Your instructor will explain the requirements for proper eye protection for your lab. State and federal laws require that eye protection—safety glasses or safety goggles—must be worn at all times in the lab. Consult your instructor for policies regarding wearing contact lenses in the lab.

Always wear appropriate clothing in the laboratory. Shoes must be closed to completely cover your feet—no sandals or flip-flops. Shirts should be long enough to cover the top of your pants. Shorts are not allowed in the lab.

Wear protective clothing as appropriate for the lab. Lab aprons or lab coats may be worn to protect your body and clothing from spills and splashes. Gloves of the appropriate type should be worn when handling dangerous or toxic chemicals.

Tie back long hair so that it does not fall into chemicals or an open flame.

Never eat or drink in the lab and do not bring opened food containers or drinks into the lab.

Prior to the beginning of each lab, remove all loose papers from the lab bench and move all book bags and purses out of your work area to an appropriate spot away from the lab benches.

Turn off and put away all electronic devices including cell phones and mp3 players before entering the laboratory.

Handling Chemicals in the Laboratory

Before dispensing any chemical, always check the label twice to make sure you have the chemical indicated in the experimental procedure. Check the label for a hazard symbol (discussed in "Waste Disposal and Cleanup" below) and handle the chemical appropriately for its hazard designation.

Information concerning the potential hazards and appropriate handling of any given chemical is provided in the Material Safety Data Sheet (MSDS). The Occupational Safety and Health Administration of the United States requires that MSDSs be made available to those who work with chemicals. A variety of websites provide access to MSDSs at no charge. Your instructor will discuss your institution's policies and the availability of the MSDSs appropriate to each of the experiments in this course.

Never put any chemicals in your mouth.

When dispensing a liquid, if some of the liquid spills on the side of the bottle or on the lab bench, clean up the spill immediately using appropriate procedures and caution.

If a liquid spills onto your clothing (but does not come into contact with your skin) immediately remove the affected clothing.

When dispensing a solid, take note of any of the solid that is spilled onto a balance or the lab bench and use appropriate procedures to clean it up. Never weigh a solid directly onto a balance; instead weigh it into a plastic weighing boat or a beaker.

When measuring a chemical, never return any excess to its original container; otherwise, you may contaminate the entire bottle. Ask your instructor how to dispose of any excess chemicals.

If you are instructed to smell a chemical, do so by holding the container (test tube, beaker, or bottle) 3–6 inches from your nose and wafting the vapor to your nose by waving your hand across the top of the container toward your face. This technique will be demonstrated by your instructor. Never bring the container directly beneath your nose to smell the chemical inside.

If the chemical you are dispensing is flammable, make sure there are no open flames nearby. When appropriate, dispense the chemical in a fume hood.

Laboratory Safety Equipment

Know the location and proper operation of all of the safety equipment (described below) in the lab.

Eye wash: An eye wash station is used when chemicals are splashed in your eyes. If a chemical is splashed in your eyes, place your eyes in the stream of water from the eye wash station, making sure you keep your eyelids open, and flood your eyes with water for a minimum of 10 minutes.

Safety shower: A safety shower is used when large amounts of chemicals are splashed on your body. If a large splash should occur, immediately remove your clothing, stand beneath the safety shower, and flood the affected skin with water for a minimum of 10 minutes.

Fire extinguisher: A fire extinguisher is not appropriate for every fire. A small fire contained in a beaker can be extinguished by covering it with a watch glass or inverting a larger beaker over the beaker containing the fire. When a fire extinguisher is needed, the nozzle of the extinguisher should be directed at the base of the fire and the trigger pulled to activate the extinguisher. Your instructor will explain the type(s) of fire extinguishers available in your laboratory.

Fire blanket: Many labs have fire blankets that can be used to wrap a person whose clothes or hair have caught on fire.

Fume hoods: A fume hood is used to dispense volatile or particularly noxious chemicals. When using a fume hood, make sure the sash on the hood is raised only to the level indicated on the hood.

Waste Disposal and Cleanup

Check with your instructor before disposing of any excess or waste chemicals. Always properly dispose of all chemicals regardless of their hazard.

Before you dispense any chemical, take note of the hazard symbol, if present, on the label. An often used chemical hazard symbol is the diamond-shaped icon of the National Fire Protection Association, often called an NFPA diamond. The diamond is divided into four areas as shown in Figure 1.1. The portion at the top of the diamond is the Flammability rating, colored red; the portion on the right is the Reactivity rating, colored yellow; the portion on the left is the Health hazard rating, colored blue; and the portion on the bottom is the Special Hazard information, colored white. The numbers in the first three areas range from 0–4 with 0 being the lowest rating, indicating no hazard, and 4 being the highest rating, indicating a serious hazard. The Special Hazard area of the diamond contains letter codes such as W̶ to indicate that the chemical is dangerous when in it comes in contact with water and OX to indicate that the chemical is an oxidizer.

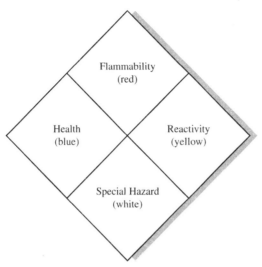

▲ FIGURE 1.1 The chemical hazard symbol.

Emergency Procedures

Immediately report any accident, no matter how minor, to your instructor. A small cut may seem insignificant to you, but you must report it. Also inform your instructor of any spills that occur at or around your work area.

Cuts and burns are the most common injuries in a lab. In the case of a small cut, flush the area with clean water and apply pressure to stop the bleeding. Make sure you or your lab partner report the cut immediately. In the case of a burn, flush the affected area with cool water and notify your instructor immediately.

Chemical splashes and splatters on your skin or in your eyes have the potential of being extremely serious and must be treated without delay. You should carefully read the preceding sections about the eye wash and the safety shower under "Laboratory Safety Equipment". Familiarize yourself thoroughly with the location and operation of these pieces of safety equipment in the laboratory.

Commitment to Laboratory Safety Pledge

I have read the previous material and understand the safety procedures I am required to follow during this laboratory course. I agree to comply with all safety rules and regulations and pledge to:

_____ Turn off and put away all electronic devices including cell phones and mp3 players before entering the laboratory.

_____ Prepare for each lab by reading and studying the experiment and all associated instructions prior to coming to the laboratory.

_____ Arrive on time to the lab and listen carefully to any safety instructions for the day's lab.

_____ Wear approved safety goggles or glasses at all times when in the laboratory.

_____ Wear appropriate and sensible clothing and closed shoes and tie back long hair.

_____ Never put anything in my mouth or eat or drink while in the laboratory.

_____ Know the location and proper operation of safety equipment, including eye wash stations, safety showers, and fire extinguishers.

_____ Clear my work area and lab bench of all items not associated with or required for the day's work prior to beginning work in the laboratory.

_____ Never work alone in the laboratory.

_____ Follow the designated experimental procedures as written and never perform unauthorized experiments.

_____ Carefully read the label on all chemicals before dispensing. Dispense small amounts of chemicals to avoid excess and dispose of any excess according to directions provided by my instructor.

_____ Clean up chemical spills immediately using appropriate procedures for the specific chemical.

_____ Dispose of all waste using appropriate procedures and the proper containers.

_____ Inform my instructor immediately of any spills or accidents in the laboratory.

_____ Carefully handle objects that have been heated so as to avoid burns.

_____ Clean up my work area and lab bench as well as all common areas and wash my hands before leaving the laboratory.

_____
Printed Name

_____
Your signature

_____   _____
Laboratory class and section   Date

EXPERIMENT

1. In the space provided on the report sheet, draw a layout of the laboratory, showing all of the lab benches, doors, windows, and fume hoods. On your map, indicated the location of all safety equipment, including eye wash stations, safety showers, fire extinguishers, fire alarms, and fire blankets. Label all exits.

2. Place a small amount of table sugar (a teaspoon or less) in a 50 mL beaker. Using a small pipette or dropper, carefully place 2 drops of concentrated sulfuric acid, H_2SO_4, directly on the sugar in the beaker. (CAUTION: Sulfuric acid is highly corrosive! Wear gloves and handle with care.) Set the beaker aside in a safe place on your bench. Record your observations on your report sheet at the end of the lab.

3. Obtain two small squares of cotton cloth and two watch glasses. Place one square of cloth on each of the watch glasses.

   a. To the first square of cloth, add 5 or 6 drops of concentrated sulfuric acid directly on the cloth. (CAUTION: Sulfuric acid is highly corrosive! Wear gloves and handle with care.) After 5 minutes, record your observations on the report sheet. Dispose of the cloth as directed by your instructor. Clean the watch glasses as directed by your instructor for before moving on to the next step.

   b. Take the second watch glass and cloth to the fume hood or other well-ventilated area as indicated by your instructor. To the second square of cloth, add 5 or 6 drops of concentrated hydrochloric acid directly on the cloth. (CAUTION: Hydrochloric acid is highly corrosive! Wear gloves and handle with care.) After 5 minutes, record your observations on the report sheet. Dispose of the cloth as directed by your instructor. Clean the watch glasses as directed by your instructor for before moving on to the next step.

4. On two clean and dry watch glasses, place a small sample of egg white.

   a. To the first sample of egg white, add 5 or 6 drops of sodium hydroxide solution. (CAUTION: Sodium hydroxide is highly caustic! Wear gloves and handle with care.) Stir the egg white and sodium hydroxide with a glass stirring rod for 1 minute. Record your observations on the report sheet. Dispose of the contents of the watch glass as directed by the instructor. Clean and dry the watch glass as directed.

   b. Take the second watch glass to the fume hood or other well-ventilated area as indicated by your instructor. To the second sample of egg white, add 5 or 6 drops of concentrated nitric acid. (CAUTION: Nitric acid is highly corrosive! Wear gloves and handle with care.) Stir the egg white and nitric acid with a glass stirring rod for 1 minute. Record your observations on the report sheet. Dispose of the contents of the watch glass as directed by the instructor. Clean and dry the watch glass as directed.

5. Place 3–5 Styrofoam (polystyrene) "packing peanuts" into a 100 mL beaker. Dampen the peanuts with a few squirts of acetone from a wash bottle. Record your observations on the report sheet.

Pre-Lab Questions | 1

Study the lab scene below and locate at least ten safety violations.

List each of the lab safety violations you found in the lab scene and write brief instructions on how the violation should be corrected.

1.

2.

3.

4.

5.

6.

7.

8.

9.

10.

REPORT SHEET | LAB

Why Is Safety So Important in Chemistry Lab?

1

1. Draw and label the map of your laboratory here.

2. Record your observations of the reaction of sugar and sulfuric acid.

3. Cotton cloth: (record your observations of each reaction)

   a. Cotton cloth and sulfuric acid

   b. Cotton cloth and hydrochloric acid

4. Egg whites: (record your observations of each reaction)

   a. Egg white and sodium hydroxide

   b. Egg white and nitric acid

5. Record your observations of adding acetone to Styrofoam packing peanuts.

QUESTIONS

1. The squares of cotton cloth were used to simulate clothing and the way clothing reacts to chemical spills or splashes.

   a. If one of the acids used in this experiment splashed on your clothing, what is the appropriate response assuming that the chemical did not come in contact with your skin?

   b. If the chemical soaked through your clothing and came in contact with your skin, what is the appropriate response?

2. Egg whites are composed mostly of protein and water. We used them to simulate what might happen to your skin or eyes, which are mostly protein, if they were splashed with a chemical. Based on your observations of the reaction of each of the chemicals with the egg whites, describe how each chemical would affect your skin and eyes.

3. The Styrofoam packing peanuts used in this experiment are made of a common plastic material, polystyrene, which is used in a variety of products including CD cases, yogurt containers, and plastic cups.

 a. List some items that are made of or contain plastic that you bring to lab with you each week.

 b. How can you make sure these items are safe in a laboratory setting?

 c. Many calculators have plastic cases and plastic parts. Considering that you may have to use your calculator in the lab, what steps will you take to make sure it is not damaged by acetone or other chemicals?

4. Sulfuric acid is a dehydrating agent that reacts with many things by removing water (H_2O) molecules. Sugar's chemical formula is $C_{12}H_{22}O_{11}$. If sulfuric acid dehydrates sugar, what is the black substance left in the beaker at the end of the lab?

How Will I Use Math and Measurement in My Career? | 2

LEARNING GOALS

After completing this laboratory exercise, you should be able to:

- Discuss the SI system of units and identify the standard and base units for mass, volume, and length
- Explain the use of prefixes in the SI system
- Describe the English system of units and list common equivalencies between units
- Demonstrate the use of conversion factors to convert from one unit to another within a system or between systems
- Demonstrate the use of conversion factors for solving different types of problems
- Explain how to determine body frame size using measurements of wrist circumference and elbow width

BEFORE YOU BEGIN

In Chapter 1 of your text (*General, Organic, and Biological Chemistry, 2nd edition*), you were introduced to a variety of mathematical concepts that are important in chemistry, especially in the laboratory setting. Before you begin this laboratory exercise, review the material in Chapter 1, focusing on the following sections and concepts:

- Section 1.3—Systems of measurement, prefixes, conversion factors, significant figures, and rounding
- Section 1.5—Accuracy, precision, and dosage calculations

Read the introduction that follows; then work through the pre-lab exercise at the end of this experiment to determine whether you understand the concepts necessary to complete this lab.

INTRODUCTION

Measurement Systems

The most widely used system of measurement is the International System of Units, which is known as the SI system, from the French *système international*. The SI system was derived from, and is still often referred to as, the metric system. The United States uses a different system of measurement known as the English system. As we learn to use measurement systems, we will explore both systems and see how they are related.

Because science is an international endeavor, scientists worldwide have adopted a single system of measurement to facilitate comparison of scientific measurements and communication of those measurements. The SI system was chosen not only because of its widespread use, but also because it employs a base ten system, which makes it more convenient for calculations and conversions, as we will see.

The SI System

The SI system uses a series of standard units for measurable quantities such as mass, volume, and length. Table 2.1 shows each of these quantities, the SI standard unit, and the accepted abbreviation.

TABLE 2.1 SI Standard Units

Measurable Quantity	Standard Unit	Abbreviation
Mass	kilogram	kg
Volume	liter	L
Length	meter	m

These units serve as the standard for each of the given quantities, but the units can sometimes be awkward for certain measurements. For example, imagine trying to determine the distance from Atlanta, Georgia, to Seattle, Washington, in meters (a meter is a little longer than a yard.) Or think about trying to measure the appropriate dose of medicine for an infant in liters (a liter is a little more than a quart.) These examples demonstrate the power and convenience of the SI system for measurements large and small. Because the SI system is a base ten system of measurement, a set of prefixes was developed and can be applied to each of the standard units to change the magnitude of the unit by a power of ten. (You are already familiar with the powers of ten concept from our system of currency, where 10 pennies equal one dime and 10 dimes equal one dollar.) For example, the distance from Atlanta to Seattle could more easily be measured in kilometers, where the prefix *kilo* indicates "a thousand times." So a kilometer is 1000 meters, and the distance from Atlanta to Seattle is approximately 3500 km instead of 3,500,000 meters! Table 2.2 shows the most useful prefixes for our purposes in the lab.

TABLE 2.2 Common Prefixes and Their Meaning

Prefix	Abbreviation	Relationship to Base Unit
mega	M	$1,000,000 \times$
kilo	k	$1000 \times$
deka	da	$10 \times$
base unit (no prefix)		$1 \times$ (gram, liter, meter)
deci	d	$\div 10$
centi	c	$\div 100$
milli	m	$\div 1000$
micro	μ	$\div 1,000,000$
nano	n	$\div 1,000,000,000$

Table 2.2 can help you understand the relationship of a prefix to the base unit. In the example used previously, we saw that a kilometer is 1000 meters. Relating this to the table, we see that the prefix *kilo* means to multiply the base unit (meter in this case) by 1000 [$1000 \times$ (as shown in Table 2.2)]. Similarly, if the unit we are working with is a centimeter (*centi* is the prefix), we would know to divide the base unit by 100 [$\div 100$ (from Table 2.2)]. In other words, a centimeter is 1/100th of a meter, or said another way, there are 100 centimeters in a meter.

The English System

Unlike the SI system, the English system does not use a set of standard units and prefixes. Instead, the English system uses many different units based on the size of what is being measured. Table 2.3 shows some of the customary units for mass, volume, and length.

TABLE 2.3 English Units

Measurable Quantity	Customary Units	Abbreviation
Mass	ounce	oz
	pound	lb
Volume	quart	qt
	gallon	gal
Length	inch	in
	foot	ft

The English system is not a base ten system of measurement, but is a system derived from a variety of historical measurements. To someone unfamiliar with the system, the units and their relationships can seem random—the same is true for many who have grown up using the English system! Because the measurements are historically derived, the only way to know them and their equivalency relationships is to memorize them. Table 2.4 shows some of the common equivalencies of measurements in the English system.

TABLE 2.4 English Units and Their Equivalencies

Quantity	Unit	Equivalency
Mass	ounce (oz)	16 oz = 1 pound (lb)
	pound (lb)	2000 lb = 1 ton
Volume	teaspoon (tsp)	3 tsp = 1 tablespoon (Tbsp)
	tablespoon (Tbsp)	4 Tbsp = ¼ cup
	fluid ounce (fl oz)	8 fl oz = 1 cup
	cup (c)	2 cups = 1 pint
	pint (pt)	2 pints = 1 quart
	quart (qt)	4 qt = 1 gallon
Length	inch (in)	12 in = 1 foot
	foot (ft)	3 ft = 1 yard
	mile (mi)	1 mile = 5280 ft (1760 yd)

Table 2.4 can help you understand the relationship between and among units. The units of volume are perhaps the most complex, but they are widely used. In cooking, we typically see units of teaspoon and tablespoon; yet we see the unit gallons when we buy gasoline for a vehicle. Recipes rarely call for a gallon of an ingredient, but no one buys tablespoons of gasoline. As these examples demonstrate, proper use of the English system comes from familiarity with the system and a great deal of practice.

Relating the Systems

Because the two systems of measurement use different units, it is often necessary to convert a measurement from one system to the other, especially in the case of international trade or common measurements such as distances in sporting events. For such instances, we will use three simple equivalencies between the systems shown in Table 2.5.

TABLE 2.5 Equivalencies for English and SI Units

Quantity	Equivalency
Mass	2.2 lb = 1 kg
Volume	1.057 qt = 1.0 L
Length	1 in = 2.54 cm

A variety of other equivalencies are commonly used, and you may find them useful. However, using these three simple equivalencies (and less memory space) and the methods discussed below, you will be able to convert measurements between the two systems.

Conversion Factors

As we mentioned in the previous sections, units of measurement are related through equivalencies. In the metric system, these equivalencies relate to the prefixes (see Table 2.2) and in the English system, the equivalencies are historical (see Table 2.4). Equivalencies are relationships between a pair of units and can be expressed mathematically. For example, in Table 2.2, we see that a decimeter is 1/10th of a meter, or 10 dm = 1 m. Mathematically, we express this relationship as two ratios:

$$\frac{1 \text{ meter}}{10 \text{ decimeter}} \quad or \quad \frac{10 \text{ decimeter}}{1 \text{ meter}}$$

The ratio on the left is read "1 meter is equivalent to 10 decimeters," and the ratio on the right is read "10 decimeters is equivalent to 1 meter." Such equivalencies are called **conversion factors**. Conversion factors give the relationship between two units and can be used to convert a measurement taken in one unit to the other unit. The following example demonstrates this use of conversion factors.

How many grams of aspirin are in a 500 mg tablet? The question asks for grams and gives milligrams (mg). In Table 2.2, we see that a milligram is 1/1000th of a gram. The relevant conversion factors are:

$$\frac{1 \text{ g}}{1000 \text{ mg}} \quad or \quad \frac{1000 \text{ mg}}{1 \text{ g}}$$

How do we know which one to use? Note that the starting measurement is in milligrams, and the question asks for the measurement in grams. When choosing a conversion factor, the unit of the starting measurement should be in the denominator of the ratio and the unit of the desired measurement should be in the numerator of the ratio.

$$\text{Correct conversion factor} = \frac{\text{Desired unit}}{\text{Starting (or given) unit}}$$

Now we set up an equation to convert from the starting unit to the desired unit by multiplying the starting quantity and unit by the conversion factor.

$$500 \text{ mg} \times \frac{1 \text{ g}}{1000 \text{ mg}}$$

Cancel the common units.

$$500 \text{ mg} \times \frac{1 \text{ g}}{1000 \text{ mg}}$$

Do the math. $(500 \times 1) \div 1000 = 0.5$

This gives the final answer with the desired units, as shown here.

$$500 \text{ mg} \times \frac{1 \text{ g}}{1000 \text{ mg}} = 0.5 \text{ g}$$

The equivalencies of the English system (see Table 2.4) can be used as conversion factors as well, as the following example illustrates.

How many inches are in 3 feet?

      Starting quantity and unit: 3 ft

      Desired unit: inches

      Conversion factor with desired unit in numerator: $\dfrac{12 \text{ in}}{1 \text{ ft}}$

      Equation: $3 \text{ ft} \times \dfrac{12 \text{ in}}{1 \text{ ft}}$

      Cancel units and do the math: $3 \text{ ft} \times \dfrac{12 \text{ in}}{1 \text{ ft}} = 36 \text{ in}$

For conversions where the units involved do not have a direct equivalency, more than one conversion factor may be required. In such cases, it is often helpful to relate both units to a common unit (in the SI system, this is most commonly the base unit). Conversions between the SI and English systems is a good example of this type of conversion. When converting between the SI and English systems, the equivalencies in Table 2.5 serve as special conversion factors that we will refer to as "translators" because they allow us to translate measurements from one system to the other. When a translator is used to convert units, all units must be related to the units in the translator. The following example illustrates this type of conversion.

The recommended daily allowance of protein for an average person is about 50 g. How many ounces of protein are in 50 g? When converting between measurement systems, we rely on the translators from Table 2.5 and relate all other units to them as follows:

Starting quantity and unit: 50 g

Desired unit: oz

From Table 2.5, we find that the translator between systems is 2.2 lb = 1 kg.

Translator: $\dfrac{2.2 \text{ lb}}{1 \text{ kg}}$

Other relevant conversion factors: $\dfrac{1 \text{ kg}}{1000 \text{ g}}$ and $\dfrac{16 \text{ oz}}{1 \text{ lb}}$

Equation: $50 \text{ g} \times \dfrac{1 \text{ kg}}{1000 \text{ g}} \times \dfrac{2.2 \text{ lb}}{1 \text{ kg}} \times \dfrac{16 \text{ oz}}{1 \text{ lb}}$

Cancel units and do the math: $50 \, \cancel{g} \times \dfrac{1 \, \cancel{kg}}{1000 \, \cancel{g}} \times \dfrac{2.2 \, \cancel{lb}}{1 \, \cancel{kg}} \times \dfrac{16 \text{ oz}}{1 \, \cancel{lb}} = 1.76 \text{ oz}$

Note that the first conversion factor was used to convert the starting unit to the same unit as the translator (g to kg). The final conversion factor was used to convert the unit from the translator to the desired unit (lb to oz).

Problem Solving with Conversion Factors

Conversion factors are an important tool for more than simply converting from one unit of measurement to another. Keeping in mind that all measured quantities have units (as well as numerical values), you can use the method demonstrated above to solve a variety of mathematical problems regardless of whether they are related to chemistry, healthcare, or daily life. The key is to find the equivalency relationship(s) between the starting or given quantity and unit and the desired unit.

Let's follow the steps to solve a problem you might encounter in a healthcare scenario.

Your son wakes up with a high fever, and you call your physician to ask what you should do. She tells you to give him 100 mg of acetaminophen. In your medicine cabinet, you have a children's liquid version of the medicine, and the label indicates that the liquid contains 500 mg of acetaminophen per 5 mL of liquid. How many milliliters should you give your son?

Following the steps above, first determine the starting quantity and unit. In this case, that would be the amount ordered by your physician:

Starting quantity and unit: 100 mg

Now determine the desired unit. In this situation, you need to know how many milliliters to administer.

Desired unit: mL

Next, you need a conversion factor. Look for a relationship between milligram and milliliter from the information you have. Note that the acetaminophen label gives you this relationship when it states that the liquid contains 500 mg of acetaminophen per 5 mL of liquid. Remember that the desired unit should be in the numerator. (Often the relationship is not given in the problem, but is something you will be expected to know, such as those in Tables 2.2 and 2.4.)

Conversion factor with desired unit in numerator: $\dfrac{5 \text{ mL}}{500 \text{ mg}}$

Set up the equation: $100 \text{ mg} \times \dfrac{5 \text{ mL}}{500 \text{ mg}}$

Cancel units and do the math: $100 \text{ mg} \times \dfrac{5 \text{ mL}}{500 \text{ mg}} = 1 \text{ mL}$

This method of problem solving, often called *the factor label method* or *dimensional analysis*, is widely applicable in a variety of situations. Master this method and use it to solve problems not only in chemistry, but also in your life and career.

EXPERIMENT: MEASURING YOUR FRAME SIZE

(Note: You will need two partners to complete the following exercise.)

Insurance companies and healthcare professionals, including nutritionists and personal trainers, often use frame size as a factor in determining a person's overall health. In this experiment, you will use two methods to determine your frame size and evaluate how the results from the two methods compare.

Height

Each of the methods you will use to determine your frame size requires that you know your height. If you know how tall you are, record that information on the data sheet provided. If you do not know how tall you are, have your lab partners determine your height (in feet and inches) using the measuring sticks provided in the lab. If you use a meterstick, be sure to convert the measurement to feet and inches.

Determining Your Frame Size

Method 1: Circumference of Wrist

1. Using a cloth or paper tape measure, measure the circumference of your wrist at the point where a watch or bracelet would normally sit. This is the smallest part of your wrist just in front of the bony protrusion on the outside of your wrist.
2. Record your measurement on the your report sheet.
3. Have each of your partners independently measure your wrist circumference and record their measurements on your report sheet.
4. Calculate the average of the measurements and use the average to determine your frame size using the information in Table 2.6.

TABLE 2.6 Frame Size Based on Wrist Circumference*

Height Women	Small Frame	Medium Frame	Large Frame
Under 5'2"	less than 5.5"	5.5"–5.75"	over 5.75"
5'2" to 5'5"	less than 6"	6"–6.25"	over 6.25"
Over 5'5"	less than 6.25"	6.25"–6.5"	over 6.5"
Men			
Over 5'5"	5.5"–6.5"	6.5"–7.5"	over 7.5"

*http://www.nlm.nih.gov/medlineplus/ency/imagepages/17182.htm

Courtesy of the National Library of Medicine/National Institute of Health.

Method 2: Width of Elbow

1. Stand with your arms by your side. Extend your arm forward so that it is horizontal and parallel to the ground with your palm facing up.
2. Bend your elbow so that your forearm forms a 90° angle with your upper arm.
3. Using your thumb and index finger, locate the two bony protrusions on either side of your elbow. See Figure 2.1.
4. Measure the distance between those two protrusions using your thumb and forefinger to determine the gap and a ruler or tape measure to measure the width of the gap. Alternatively, you can use a caliper to measure the distance between the bony protrusions.
5. Record your measurement on the your report sheet.

6. Have each of your partners independently measure the width of your elbow (using the same methods) and record their measurements on your report sheet.
7. Calculate the average of the measurements and use the average to determine your frame size using Table 2.7.

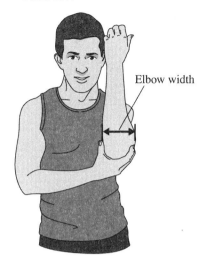

Elbow width

◀ **FIGURE 2.1** Measuring the width of your elbow.

TABLE 2.7 Frame Size Based on Elbow Width (measured in cm)*

Age Women	Small Frame	Medium Frame	Large Frame
18–24	5.6 or less	between 5.6–6.5	6.5 or greater
25–34	5.7 or less	between 5.7–6.8	6.8 or greater
35–44	5.7 or less	between 5.7–7.1	7.1 or greater
over 45	5.8 or less	between 5.8–7.2	7.2 or greater
Men			
18–24	6.6 or less	between 6.6–7.7	7.7 or greater
25–34	6.7 or less	between 6.7–7.9	7.9 or greater
35–44	6.7 or less	between 6.7–8.0	8.0 or greater
over 45	6.7 or less	between 6.7–8.1	8.1 or greater

*Adapted from *American Journal of Clinical Nutrition*, 1984, 40: 808–819

Pre-Lab Questions | 2

1. Determine the number of significant figures in each of the following measurements. Refer to the Appendix or your text to review significant figures.

   a. 0.390 g

   b. 5.04 cm

   c. 0.0612 mL

   d. 20,000 lb

   e. 407.0 L

   f. 0.81070 m

2. Complete each of the following calculations and round your answer to the correct number of significant figures. Refer to the Appendix or your text for information on calculations and significant figures.

   a. $12.3 + 0.357$ b. 9.0×2.54

   c. $96{,}456 - 821.7$ d. $0.435 \div 1.38$

3. Many pharmaceuticals are dispensed in elixirs with the medicine dissolved in a pleasant tasting liquid. This is especially true of medicines for children. The label on one such elixir reads as follows: 100 mg acetaminophen in 5 mL of liquid.

   a. Write a conversion factor showing the amount of acetaminophen in the liquid.

   b. Your doctor prescribes 200 mg of acetaminophen for your child's fever. How many milliliters of the liquid described above should you give your child? (Use the conversion factor you wrote in part a.)

4. Note the position of the arrows on each of the targets below and label each pattern as high accuracy/low precision, high precision/low accuracy, low precision/low accuracy, or high accuracy/high precision.

Name _____

Date _____ Lab Section _____

How Will I Use Math and Measurement in My Career?

Practice with Conversion Factors and Problem Solving

1. Complete each of these conversions using the equivalencies in Tables 2.2 and 2.4. Show your work in the space provided.

   a. 21 dag = _____ g

   b. 15 tsp = _____ Tbsp

   c. 0.45 m = _____ cm

   d. 33 ft = _____ yd

   e. 150 mL = _____ L

   f. 1500 lb = _____ tons

2. Complete each of the conversions below. Tables 2.2 and 2.4 do not have direct equivalencies for most of these conversions, so you may need more than one conversion factor. Show your work in the space provided.

   a. How many milligrams of vitamin C are in a tablet that contains 10,000 micrograms of vitamin C?

   b. How many cups of milk are in a 1-gallon container of milk?

   c. A newborn baby weighs 7.5 lb and is 1.75 feet long. What is the baby's weight in ounces and length in inches?

   d. How many deciliters of soft drink are in a bottle containing 500 mL?

3. Use the translators from Table 2.5 and the appropriate equivalencies from Tables 2.2 and 2.4 to complete each of the following conversions.

   a. How many milliliters of coffee are in 1 cup?

   b. Using the information in problem 2c, calculate the baby's weight in kilograms and length in centimeters.

   c. At your annual physical, your doctor tells you that you need to lose 5 kg. Your bathroom scale only measures in pounds. How many pounds do you need to lose?

   d. During a surgical procedure, a patient receives 2.5 pt of blood. How many liters of blood did the patient receive?

4. Your chemistry professor entered a 5K race and was among the top ten finishers. Your biology professor entered the same race, but dropped out at the 3K mark, too winded to finish. Given that a 5K covers a distance of 5 km and that 1 mi = 1.61 km, answer the following questions.

   a. How many miles did your chemistry professor cover in the race?

   b. How many miles did your biology professor cover in the race?

   c. Show the conversion of kilometers to miles using the information in Tables 2.2, 2.4, and 2.5.

5. A kitchen measuring spoon has a label that reads "1 tsp, 5 mL." Using the equivalencies in Tables 2.2, 2.4, and 2.5, determine how many milliliters are in 1 teaspoon and decide if the label on the spoon is correct.

6. Pharmacists sometimes measure medicines in the unit of "grains," where 1 gr = 65 mg. The label on a bottle of aspirin reads "Aspirin, 5 gr." How many mg of aspirin are in a tablet containing 5 gr?

7. From goal line to goal line, a football field is 300 ft long. If a player catches the ball while standing on one goal line, runs to the other goal line, and scores a touchdown, how many meters did he run?

8. Medications are often prescribed in dosages of milligrams of medication/kilogram of body mass (weight). A doctor writes a prescription for a medication at a dosage of 5 mg/kg.

   a. How many milligrams will a person who weighs 60 kg need to take?

   b. If the medication is dispensed as a liquid containing 100 mg of the medication per 1 mL of the liquid, how many milliliters of the liquid will the person need to take?

Experimental Data and Results: Measuring Your Frame Size

Your height _____

Wrist Circumference

Measurements

Yours (in)	Partner 1 (in)	Partner 2 (in)	Average (in)

Frame size _____

Elbow Width

Measurements

Yours (in)	Partner 1 (in)	Partner 2 (in)	Average (in)

Frame size _____

QUESTIONS

1. Compare the three measurements of your wrist circumference.

    a. Are the measurements exactly the same? If not, why are they different?

    b. Would you consider the measurements to be similar (precise)?

2. Compare the three measurements of your elbow width.

    a. Are the measurements exactly the same? If not, why are they different?

    b. Would you consider the measurements to be similar (precise)?

3. For each measurement of your elbow, the instructions indicated that each of your partners was to do the measurement "independently," meaning that each person did the measurement without looking at the measurements taken by the others. Why was this important?

4. Why do scientists typically take multiple measurements of the same thing, as you did with your wrist and elbow?

5. Did both measurements give the same frame size for you?

Why Is Measurement Important in Chemistry? | 3

LEARNING GOALS

After completing this laboratory exercise, you should be able to:

- Describe the significance and use of measurement in chemistry
- Summarize the scientific method of inquiry
- Explain how to measure a substance's mass and volume and discuss the difference between mass and weight
- Determine a substance's density and explain how density can be used to analyze the purity and/or identity of the substance

BEFORE YOU BEGIN

In Chapter 1 of your text (*General, Organic, and Biological Chemistry, 2nd edition*), you were introduced to matter and some properties of matter. Before you begin this laboratory exercise, review the material in Chapter 1 focusing on the following sections and concepts:

- Section 1.4: Properties of matter—mass, volume, and density

Read the introduction that follows; then work through the pre-lab exercise at the end of this experiment to determine whether you understand the concepts necessary to complete this lab.

INTRODUCTION

Scientific Inquiry and Measurement

At its very core, science is not a collection of facts, but is a way of observing and trying to understand our world. Observation is the key to science and scientific inquiry because observations lead to questions about what we see. These questions lead to possible explanations, and such explanations often lead to experimentation. This series of observation, question, possible explanation, and experimentation compose the scientific method of inquiry. It is the method that scientists use to learn about our world.

When studying chemistry, especially in the laboratory, we do quite a bit of measuring as part of our inquiry. For example, we measure mass, weight, volume, and length. But why all of this measuring? Measurement seems more appropriate for math or cooking.

As we discussed in Chapter 1, chemistry is the study of the properties and behavior of matter. Measuring the mass or volume of a sample of matter helps us understand its properties. In turn, the properties of matter can help us identify what it is and how we might take advantage of its properties. For example, by measuring and comparing the properties of a sample of cotton versus a sample of aluminum, we can determine that the cotton would be more appropriate for dressing and caring for a wound than would the aluminum.

In this lab, we will introduce some fundamental properties of matter, learn to measure and analyze those properties, and bring all of this together to help answer a "currency" question.

Measurable Physical Properties

In the textbook, matter was defined as anything that takes up space and weighs something. Another way of stating this is that matter is anything that has volume and mass. Both mass and volume are properties that all matter possesses. Let's consider these two properties and the way they can be measured.

Mass is the measure of the amount of matter in a substance. Mass is measured in the lab using a balance similar to the one shown in Figure 3.1. When you think of measuring a substance on a balance or scale, you most likely think of determining the substance's weight. Practically, this is true, but there is a difference between mass and weight. Weight changes based on the pull of gravity on an object, but mass does not depend on location and therefore does not change. So a man standing on the surface of the moon weighs less than he did when standing on Earth because the moon exerts less of a gravitational pull on him. The amount of matter in the man did not change. (He is not less of a man when he is on the moon.) Because the gravitational pull on a substance is relatively uniform across the surface of the earth, the mass and weight of a substance will have the same measured value.

In a scientific laboratory, mass is typically measured in the unit of grams (g). A single raisin or paper clip has a mass of approximately 1 g.

▲ **FIGURE 3.1** A typical laboratory top-loading balance. A balance is used to determine the mass of a substance.

Volume is the measure of the amount of three-dimensional space a substance occupies. In the lab, volume can be measured using a graduated cylinder similar to the one shown in Figure 3.2, which uses the unit of milliliters (mL). In a clinical setting, volume is often measured using a syringe, which gives the volume in cc's, or cubic centimeters (cm^3). The units are actually equivalent because 1 milliliter equals 1 cubic centimeter.

Graduated cylinders and syringes are useful for measuring the volume of liquids, but what if you need to know the volume of a solid? For a regularly shaped solid such as a cube, you could measure its height, width, and depth and then multiply those measurements to get the volume of the solid. However, a simpler method is to submerge the solid in a graduated cylinder containing a known volume of liquid and record the difference in volume before and after the solid is submerged. The difference in the volume is the amount of liquid displaced by the solid and is equal to the volume of the solid.

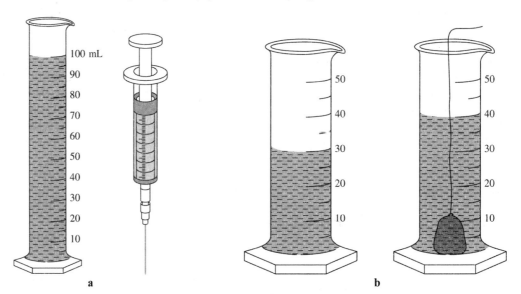

▲ **FIGURE 3.2** (a) A graduated cylinder and syringe can both be used to easily determine the volume of a liquid. (b) A graduated cylinder can also be used to determine the volume of a solid by fluid displacement.

The final property of matter we will consider in this experiment is density. Density is the ratio of a substance's mass to its volume and can be expressed mathematically as shown. The typical unit for density is g/mL for liquids and solids.

$$\text{Density} = \frac{m}{V} \text{ (where m is mass in grams and V is volume in milliliters)}$$

Density can be thought of as a measure of the degree of packing of matter and is a characteristic physical property of a substance. In other words, density can help determine the identity of a substance. For example, if someone gives you a piece of metal and says that it is "pure gold," you can measure its density to determine whether the sample is in fact pure. Pure gold metal has a density of 19.3 g/mL. If mixed with other metals, the density of a sample of the exact same size will usually decrease, indicating that the sample is not pure gold.

Density can be used to determine a person's percent age of body fat by a method called hydrostatic weighing—weighing a person underwater. Because fat is less dense than bone or muscle, a person with a higher percentage of body fat tends to float in water and, therefore, weighs less underwater than a person with less body fat. Hydrostatic weighing requires special equipment and people trained to use the equipment, but it is considered the most accurate method for determining body fat percentage.

EXPERIMENT: WHAT HAPPENS TO THE MASS OF A PENNY AS IT CIRCULATES?

The official website of the United States Mint (http://www.usmint.gov) indicates that the penny is still the most widely used denomination of American currency still in circulation. Such heavy usage means that pennies are handled, carried in pockets and purses, dropped and retrieved, thrown in wishing wells, left in cups at convenience stores, picked up from parking lots, etc. How does all of that use affect the mass of a penny? Does a penny collect dirt and grime as it circulates? Does it rub against other coins and lose some of its surface? In this experiment, you will use the method of scientific inquiry to investigate pennies of various ages to attempt to discover what happens to the mass of a penny as it circulates.

Part 1

Hypothesis

To begin the process of scientific inquiry, scientists propose an educated guess indicating what they think they may discover during their investigation. This educated guess is known as a **hypothesis**. A hypothesis guides the formulation of questions and experiments that are designed to test its validity.

Develop your hypothesis with regard to this question: What happens to the mass of a penny as it circulates? *Record your hypothesis on the report sheet.*

Collecting Data

To begin the investigation, you will use a top-loading balance to determine the mass of pennies from a variety of years.

1. Starting with pennies from the mid-1960s and moving forward to the present, collect 3–5 pennies for a specific year within a five-year period. For example, you could collect four pennies from 1967 to represent the 1965–1969 time frame and five pennies from 1974 to represent the 1970–1974 time frame.
2. Weigh 3–5 pennies from one of the years you have collected to within 0.01 gram. Calculate and record *in Table 3.1 on your report sheet* the average mass of a penny from that year and the year of its mint. (Keep your pennies for the next part of the experiment.)
3. Repeat step 2 for the pennies you have collected from each year.
4. Share your data with your instructor, who will compile the data for the entire class and create a table similar to Table 3.1 showing the compiled class data.
5. Using the table of compiled class data, prepare a graph with *Year of Mint* along the x-axis and *Average Mass of Penny* along the y-axis. Follow the "Guidelines for Graphs" to prepare your graph of the data.

Guidelines for Graphs

A proper graph must:

- Take up most of the area of the sheet of paper.
- Include a title indicating what data the graph represents.
- Include clear, definitive labels on each axis, including units where appropriate.
- Give the name(s) of the experimenter(s).

Part 2

Further Investigation

Based on the data you collected, you most likely notice a significant difference in the mass of a pre-1980 penny and a post-1985 penny. Look more carefully at the data for the 1980–1985 pennies. Is the change gradual or sudden? What is the cause of the change? Did the size of the penny change? Was there a change in the composition of the penny?

Once again, formulate a hypothesis to guide your inquiry and write that hypothesis in the space provided on the report sheet.

Collecting Data

To help you understand the change that occurred with the penny in the early 1980s, you may find it helpful to know whether the size of the penny changed. You can do a simple visual inspection as a first approximation. *Record your observation on the report sheet.*

To accurately test whether the size of the penny has changed, you can measure its volume. Because a penny is a small solid object, you can most easily measure its volume by liquid displacement as discussed earlier and as demonstrated in Figure 3.2b. However, a single penny would produce a very small change in the volume of a liquid in a graduated cylinder. Therefore, as you did with the mass, you will measure the volume of 3–5 pennies and take the average to determine the volume of a single penny. For this part of the experiment, use the same pennies you used for determining the mass.

1. Pour 30–40 mL of water into a 100 mL graduated cylinder and accurately record in Table 3.2 on the report sheet the volume to within 0.1 mL. See "Guidelines for Reading a Graduated Cylinder."
2. Place 3–5 pennies from a given year into the graduated cylinder, taking care not to splash the water out of the cylinder.
3. Read and record the new volume *in Table 3.2 on the report sheet.*
4. Calculate and record the average volume of a penny from that year *in Table 3.2 on the report sheet.*
5. Repeat steps 1–4 for the pennies you collected from each year.
6. Share your data with your instructor, who will compile the data for the entire class and create a table similar to Table 3.2 showing the compiled class data.

Guidelines for Reading a Graduated Cylinder

To accurately determine the volume of liquid using a graduated cylinder, pour the liquid into the cylinder and set the cylinder on a flat, level surface. Bringing your eyes even with the liquid level, you will notice that the liquid forms a concave surface in the cylinder. This curve is called the meniscus and is read at its lowest point, as shown in Figure 3.3.

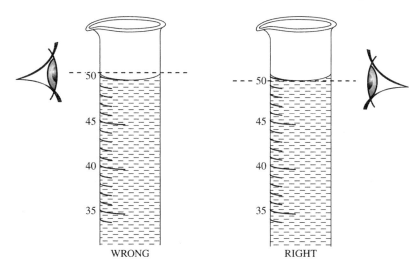

▲ **FIGURE 3.3** Reading a meniscus. Your eye should be even with the liquid level. The correct volume is obtained by reading the volume at the lowest point of the meniscus. In this case, the correct volume is 50.0 mL, not 51.0 mL.

Part 3

In the earlier section "Measureable Physical Properties," we discussed the fact that density is a characteristic property of a substance. The color of the metal used to make pennies leads us to believe that pennies are made of copper. Copper has a density of approximately 8.9 g/mL. Using the compiled data for mass of a penny and the compiled data for volume of a penny, *complete Table 3.2 in the report sheet* to determine the density of a penny from each year. Focus on the years of mint from 1975–1985.

Pre-Lab Questions | 3

1. List the "steps" of the scientific method of inquiry.

2. Define the terms *mass* and *weight*. Explain the difference between the two.

3. In a laboratory setting, volume is measured in milliliters, but it is measured in cc's (cm^3) in a clinical setting. How are a milliliter and a cubic centimeter related?

4. Density is described as a "characteristic property of matter." In your own words, explain what this description means.

Name _____

Date _____ Lab Section _____

Why Is Measurement Important in Chemistry?

Part 1

What happens to the mass of a penny as it circulates? Write your hypothesis here.

TABLE 3.1 Year of Mint versus Average Mass of Penny

Year of Mint	Average Mass of Penny (g)

1. Based on the graph of the compiled data, what can you determine about the mass of a penny as it circulates?

2. Do the data and graph support or contradict your hypothesis? Explain.

3. Make note of any unusual or unexpected results represented by the data or the graph.

4. Using your graph, determine the average mass of a pre-1980 penny and a post-1985 penny.

   Average mass of pre-1980 penny _____

   Average mass of post-1985 penny _____

37

Part 2

What is the cause of the change in the mass of a pre-1980 penny versus a post-1985 penny? Write your hypothesis here.

Does the size of a pre-1980 penny appear different than that of a post-1985 penny? Write your observations here.

TABLE 3.2 Determining Average Volume of a Single Penny from Different Years

Year of Mint	Initial Volume of Water (mL)	Volume of Water + Pennies (mL)	Volume of All Pennies (mL)	Average Volume of a Single Penny (mL)

1. Based on the table of compiled data for the class, what can you determine about the size of a penny over the years? Has it changed?

2. Do the data support or contradict your hypothesis? Explain.

3. Do the data from determining volume support or contradict your visual inspection of the pennies? Explain.

TABLE 3.3 Determining the Density of Pennies from Different Years

Year of Mint	Average Mass (g)	Average Volume (mL)	Density (g/mL)

1. Based on your determination of density recorded in Table 3.3, are all pennies made of copper? Explain.

2. Do the data support or contradict your hypothesis concerning the change in pennies that occurred during the 1980–1985 time period? Explain.

3. Based on all of the experimental data you have collected:

   a. How would you respond to the initial question, "What happens to the mass of a penny as it circulates?"

   b. How would you respond to the second question concerning the change in the mass of a penny in the time period from 1980–1985?

4. Propose further experiments you could conduct to gain further insight into the change in pennies that occurred in the early 1980s.

How Much Fat Is
in Your Milk? | 4

LEARNING GOALS

After completing this laboratory exercise, you should be able to:

- Explain how mixtures differ from pure substances
- Describe how centrifugation can be used to separate certain mixtures
- Outline a method to determine the percent fat by mass in a sample of milk

BEFORE YOU BEGIN

In Chapter 1 of your text (*General, Organic, and Biological Chemistry, 2nd edition*), you were introduced to matter and its classification as mixtures and pure substances. Before you begin this experiment, review the material indicated here focusing on the following sections and concepts:

- Section 1.1: the difference between a mixture and a pure substance and the characteristics of homogeneous and heterogeneous mixtures

INTRODUCTION

Matter can be divided into two broad categories—pure substances and mixtures. In this experiment, we will investigate a mixture with which you are likely familiar: milk. Milk is a homogeneous mixture of a variety of substances including fats, proteins, and sugars in water. In other words, any small sample of milk from a larger container has the same amount of each of the components as any other sample. We can tell this because one glass of milk from a carton tastes the same and has the same "mouth feel" as any other glass of milk from the same carton. Milk is a special type of mixture called a colloid. The fat particles in milk are small enough to remain evenly dispersed throughout the milk without settling out. A mixture that contains larger particles that settle out upon standing (such as fine silt/dirt in water) is called a suspension.

Mixtures differ from pure substances in that a mixture can be separated into its components by physical means (filtering, boiling, etc.) while a pure substance cannot. For example, a sample of salt water can be separated by boiling off and collecting the water to leave the solid salt behind. The collected water and the remaining salt, which were separated and isolated, are now pure substances.

In this experiment, we will use a physical method called centrifugation to separate some of the components of milk. A centrifuge (see Figure 4.1) is an instrument that spins tubes containing samples of a substance at high revolutions per minute (rpm). The spinning produces a force, called a centrifugal force, that will cause heavy or more dense substances to move to the bottom of the tube, while lighter or less dense substances will remain at the top. Hospitals and blood banks often use centrifuges to separate blood, also a mixture, into its components. When blood is centrifuged, the red blood cells move to the bottom and the straw-colored liquid called plasma remains on top. These two components can then be used for different medical applications.

The fat particles in milk are less dense than the water-soluble proteins and sugars. When milk is centrifuged, the fat remains on top of the sample while the water layer with its dissolved proteins and sugars moves to the bottom. The force generated by the centrifuge allows us to separate the fat in milk from the other components. We can then isolate the fat from the rest of the sample, measure the amount of fat, and determine the percent fat by mass in the milk.

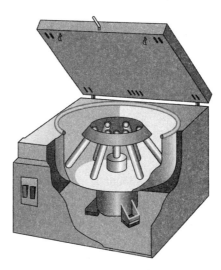

▲ **FIGURE 4.1** A typical laboratory centrifuge.

EXPERIMENT

Part 1. Grease Spot Test

When applied to plain paper such as filter paper or a brown paper bag, substances that contain fat will leave a translucent spot on the paper after drying.

1. Place 1–2 mL samples of whole milk, 2% milk, and skim milk each in a separate test tube.

2. Obtain a large piece of filter paper. Using a pencil, at the top of the circle (12 o'clock position), write "Whole," write "2%" at the 4 o'clock position, and write "Skim" at the 8 o'clock position.

3. Using a disposable pipette, place 3 drops of whole milk near the spot labeled "Whole." Add the drops slowly one at a time so that the spot that forms remains about the size of a penny or a nickel (½–¾ of an inch across).

4. Repeat step 3 with a clean disposable pipette and the 2% milk near the spot labeled "2%" and with another clean disposable pipette and the skim milk near the spot labeled "Skim."

5. Once all of the samples have been added, set the filter paper aside to dry thoroughly in a place where it will not be disturbed while you complete Part 2 of the experiment. (Alternatively, dry the filter paper carefully with a heat gun or in a lab oven. Take care not to overheat the samples.)

6. Once the samples are completely dry, visually inspect each spot. Do you notice anything on the filter paper? Hold the paper up to a light source and take note of each spot.

7. Record your observations on your report sheet.

Part 2. Separation of Fat from Milk

1. Obtain two clean centrifuge tubes. Using a permanent marker, label one "Whole" and the other "2%."

2. Determine the mass of each tube and record the masses on your report sheet.

3. Fill the centrifuge tube labeled "Whole" one-half to two-thirds full with whole milk.

4. Fill the centrifuge tube labeled "2%" one-half to two-thirds full with 2% milk.

5. Determine the mass of each tube. If their masses are not equal, add or remove milk dropwise until both tubes + milk samples have the same mass.

6. Record the mass of each tube + milk on your report sheet.

7. Place your tubes opposite each other in the centrifuge.

8. Close the cover on the centrifuge, turn it on, and set it to run for a minimum of 20 minutes.

    Note: Ensure that the centrifuge is balanced. An unbalanced centrifuge can be dangerous. You instructor will demonstrate the correct use of a centrifuge. Have your instructor check your centrifuge before you turn it on.

9. After 20–30 minutes of centrifuging, turn the centrifuge off and allow it to stop spinning on its own. DO NOT attempt to stop the centrifuge with your hand or any other object as this may cause injury and/or disrupt your samples. If the centrifuge has a lid, keep it closed until the centrifuge stops spinning.

10. Carefully remove the tubes from the centrifuge and place them in a tube stand or beaker.

11. Visually inspect each tube and record your observations on your report sheet. Do you notice any difference in the sample now as compared to before centrifuging?

12. Save both samples for Part 3.

Part 3. Determination of Amount of Fat in Milk

1. Obtain two small pieces of filter paper. Using a pencil, label one "Whole" and the other "2%."

2. Determine the mass of each piece of filter paper separately and record the two masses on your report sheet.

3. Working with the sample in the tube labeled "Whole," use a microspatula to carefully remove the layer of fat from the top of the sample. You can most easily accomplish this by holding the tube at a slight angle while inserting the tip of the microspatula between the wall of the tube and the fat layer. Carefully work the tip of the microspatula around the circumference of the fat layer to dislodge it from the sides of the tube as shown.

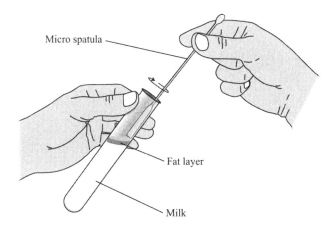

4. Once the fat layer is dislodged from the side of the tube, use the microspatula to remove it from the tube. Place the sample of fat onto the filter paper labeled "Whole."

5. Take care to wipe the entire fat sample from the microspatula onto the filter paper.

6. If the filter paper is wet with any of the milk from the sample, allow it to dry thoroughly.

7. When dry, weigh the filter paper + fat sample and record the mass on your report sheet.

8. Repeat steps 3–7 using the tube labeled "2%."

9. Save the remaining liquid in the centrifuge tube to help you answer question 2 at the end of your report sheet.

10. Complete the calculations as shown under Part 3 of your report sheet.

Pre-Lab Questions | 4

1. Explain the difference between a mixture and a pure substance.

2. What are the two types of pure substances?

3. What is the key characteristic that distinguishes a homogeneous mixture from a heterogeneous mixture?

4. A sample of a watery unknown mixture is centrifuged at high rpm for 20 minutes. After centrifugation, the sample has separated into two equal layers. Is the layer on top more or less dense than the layer on the bottom? Explain how you can be certain of this.

Name _____

Date _____ Lab Section _____

How Much Fat Is in Your Milk?

4

Part 1. Grease Spot Test

In the space below, record your observations of each of the spots on the filter paper. Consider the following questions to guide your observations. Do all of the spots look the same when held up to the light? Are any of the spots significantly different from the others? Do you notice any differences between the whole milk and 2% spots? How does the skim milk spot compare to the others?

Observations

Whole Milk Spot

2% Milk Spot

Skim Milk Spot

Part 2. Separation of Fat from Milk

1. Mass of tube labeled "Whole" _____ g

2. Mass of tube labeled "2%" _____ g

3. Mass of tube labeled "Whole" + milk _____ g*

4. Mass of tube labeled "2%" + milk _____ g*

*These two masses should be approximately equal.

5. Mass of whole milk sample (Line 3 – Line 1) _____ g

6. Mass of 2% milk sample (Line 4 – Line 2) _____ g

Observations

Centrifuged whole milk sample

Centrifuged 2% milk sample

Part 3. Determination of Amount of Fat in Milk

1. Mass of filter paper labeled "Whole"   _____ g

2. Mass of filter paper labeled "2%"   _____ g

3. Mass of filter paper labeled "Whole" + fat   _____ g

4. Mass of filter paper labeled "2%" + fat   _____ g

5. Mass of fat isolated from whole milk (Line 3 – Line 1)   _____ g

6. Mass of fat isolated from 2% milk (Line 4 – Line 2)   _____ g

Calculate the percent fat in your milk samples using the mass of fat from Part 3 and the mass of the milk sample from Part 2.

$$(\text{Mass of fat/Mass of milk sample}) \times 100 = \text{Percent fat}$$

Percent fat in whole milk   _____ %

Percent fat in 2% milk   _____ %

QUESTIONS

1. Based on the results of the grease spot test, would you classify skim milk as "fat free?" Why or why not?

2. Centrifuging raw milk (straight from the cow) to remove the fat is one method of producing skim milk. With this in mind, compare the liquid remaining in the centrifuge tube labeled "Whole" with an original sample of the skim milk you used in Part 1. Do the two liquids look identical? Hold them up to a strong light source and compare the two. Do you note any differences? Would you characterize the centrifuged milk as skim milk? Why or why not?

3. Based on your determination, what is the percent fat in whole milk? Check with several other students in your lab section to see how closely your numbers agree.

4. The U.S. Food and Drug Administration (FDA) guidelines state that whole milk contains at least 3.25% fat. Compare your findings with the FDA guidelines. Discuss any discrepancies.

5. Based on your determination, what is the percent fat in your 2% milk?

Molecular Models, Isomerism, and the Shape of Molecules | 5

LEARNING GOALS

After completing this laboratory exercise, you should be able to:

- Draw a Lewis structure, a condensed structural formula, and a skeletal structure for a simple alkane, haloalkane, and cycloalkane

- Determine the VSEPR form and geometry about a given atom of a molecule

- Explain the effect of a nonbonded pair of electrons on the bond angle and geometry of a molecule

- Identify and differentiate conformational isomers, structural isomers, and stereoisomers

BEFORE YOU BEGIN

In Chapter 3 of your text (*General, Organic, and Biological Chemistry, 2nd edition*), you were introduced to the octet rule and the covalent bonding patterns common to main-group elements such as carbon, nitrogen, and oxygen. You also learned how to draw Lewis structures of covalent compounds and how to use these structures to predict the shape of a molecule. In Chapter 4, you saw these same concepts applied to organic (carbon-containing) compounds. You were also introduced to the nomenclature of organic compounds and the variety of types of isomers that organic compounds can form. Before you begin this laboratory exercise, review the material in Chapters 3 and 4 focusing on the following sections and concepts:

- Section 3.4: Covalent bonding, preferred bonding patterns of main-group elements, and drawing Lewis structures of covalent compounds

- Section 3.6: Using VSEPR to predict molecular shape, especially of carbon-containing compounds, and nonbonding electron pairs and how they affect molecular shape

- Section 4.1: Alkane and cycloalkane structure and Lewis structures of alkanes and cycloalkanes

- Section 4.2: Condensed structures of alkanes and skeletal structures of alkanes and cycloalkanes

- Section 4.4: Nomenclature of simple alkanes, haloalkanes, and cycloalkanes

- Section 4.5: Structural and conformational isomers of alkanes, cis–trans isomers of cycloalkanes, wedge-and-dash structures, and stereoisomers

Read the introduction that follows; then work through the pre-lab exercise at the end of this experiment to determine whether you understand the concepts necessary to complete this lab.

INTRODUCTION

Atoms with eight electrons (an octet) in their valence shell, such as the noble gases of Group 8A, exhibit an unusual stability as demonstrated by their failure to react with other atoms under all but the most extreme conditions. Atoms that lack this valence octet react with other atoms to achieve this octet and the stability that comes with it. The nonmetals in Groups 1A and 4A–7A achieve a valence octet by sharing electrons with other nonmetals to form covalent bonds. When atoms share electrons to form covalent bonds, the new chemical entity formed is known as a molecule, which is more stable than the atoms are by themselves.

TABLE 5.1 Preferred Covalent Bonding Patterns of the Elements of Life

Group 1A	Group 4A	Group 5A	Group 6A	Group 7A
H—				
	—C̈—	—N̈—	—Ö—	:F̈—
		—N̈=	Ö=	
	C=	:N≡		
	=C=			
	—C≡			
			—S̈—	:C̈l—
				:B̈r—
				:Ï—

The elements of life—carbon, hydrogen, oxygen, and nitrogen—adopt specific preferred bonding patterns to achieve an octet. These patterns are shown in Table 5.1. We can use the preferred bonding patterns of these elements to draw Lewis structures when we are given the molecular formula of a compound. For example, given the molecular formula for propane, C_3H_8, we can use the preferred bonding patterns to draw the Lewis structure for the compound as shown in Figure 5.1.

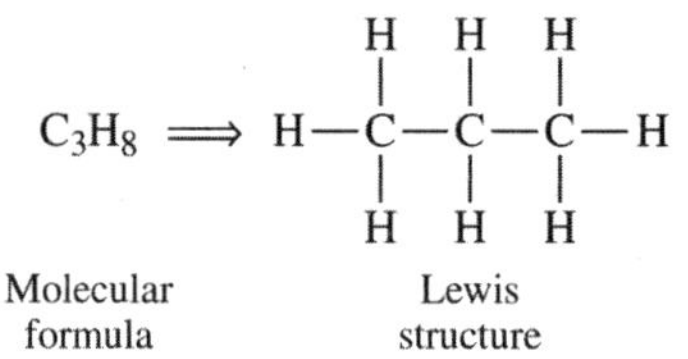

$$C_3H_8 \implies \begin{array}{ccc} H & H & H \\ | & | & | \\ H-C-C-C-H \\ | & | & | \\ H & H & H \end{array}$$

Molecular formula   Lewis structure

▲ **FIGURE 5.1** Using preferred bonding patterns to draw a Lewis structure for propane.

While the molecular formula gives the number and type of each atom in a molecule, the Lewis structure provides additional information by showing how the atoms are connected to each other. Organic compounds such as propane can also be represented by condensed structures and skeletal structures as shown in Figure 5.2. Review each of these types of structures and the wedge-and-dash structures before beginning the lab.

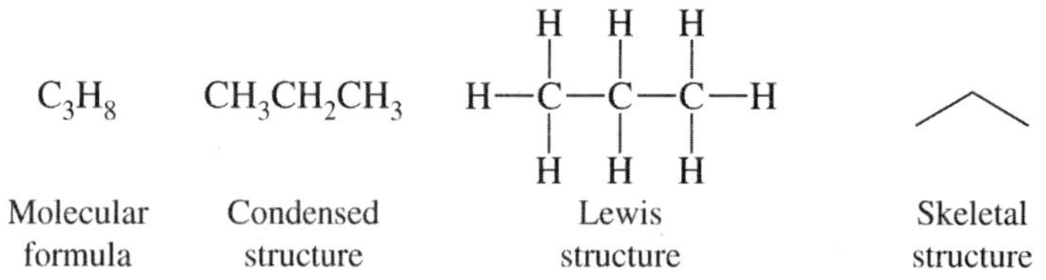

$$C_3H_8 \qquad CH_3CH_2CH_3 \qquad \begin{array}{ccc} H & H & H \\ | & | & | \\ H-C-C-C-H \\ | & | & | \\ H & H & H \end{array}$$

Molecular   Condensed   Lewis   Skeletal
formula   structure   structure   structure

▲ **FIGURE 5.2** Molecular formula and structural formulas of propane.

The Lewis structure of a molecule simply represents the connectivity of the atoms, not the shape of the molecule. However, we can use the Lewis structure to help determine the shape around an atom of a given molecule using the principles of Valence Shell Electron Pair Repulsion theory (VSEPR). To accomplish this, we will determine the VSEPR form about the atom and use that form along with Table 5.2 to determine the shape of the molecule around that atom. To determine the VSEPR form, we use the ABN shorthand: A represents the atom around which we are determining the shape; B represents the bonded electron clouds; and N represents the nonbonded, or lone, electron clouds. (Remember that double and triple bonds are counted as one bonded electron cloud.)

TABLE 5.2 Predicting Molecular Shape Using VSEPR Form

VSEPR Form	Molecular Shape	Bond Angle
AB_4	Tetrahedral	109.5°
AB_3N	Pyramidal	<109.5°
AB_2N_2	Bent	<109.5°
AB_3	Trigonal planar	120°
AB_2		180°

Let's use propane, C_3H_8, from our previous example. We will use the carbon atom on the left of the structure as the atom around which we want to know the shape (A). This carbon atom has four bonded clouds (B) due to its four single bonds and no nonbonded clouds (N). Therefore, its VSEPR form is AB_4. Comparing that to Table 5.2, we can see that the shape of the molecule around the carbon on the left is tetrahedral.

In this lab, you will get some hands-on experience with molecular structure and shape by building molecular models of organic compounds. You will use models to help you understand the connectivity of atoms in a molecule and to investigate the different types of isomerism present in organic compounds.

As you discovered in Chapter 4, isomers of organic compounds are compounds with the same molecular formula but a different connectivity or arrangement of atoms. In this lab, you will use models to see that conformational isomers are really the same compound, but with the atoms arranged differently in space by rotations around a single bond. You will compare conformational isomers to structural isomers, which have a different connectivity of atoms and, therefore, are not the same compound. Finally, you will build models of some stereoisomers and show that they are different compounds because they cannot be converted in to each other by rotation about single bonds.

EXPERIMENT: BUILDING MOLECULAR MODELS

For each part of this experiment, build the indicated models and answer each question on the report sheet for that particular model.

Part 1. Molecular Models of Simple Alkanes

1. Using the organic kit model provided, build a model of methane, CH_4. Using a wedge-and-dash structure, draw a three-dimensional model of methane on your report sheet. Draw a Lewis structure and a condensed structural formula for methane on your report sheet.
2. Build models of ethane (C_2H_6) and propane (C_3H_8). On your report sheet, draw the Lewis structure and condensed structure for each of these molecules.

Part 2. Conformational and Structural Isomers

1. Using the organic model kit provided, build a model of butane, C_4H_{10}. Arrange the carbon atoms of the model so that they are in a straight line (when viewed from above). Draw a skeletal structure for butane based on the model you built.
2. Holding one of the end carbons fixed, rotate the other end carbon 180˚ from its current position. (The rotation should occur around the bond between the second and third carbons). On your report sheet, draw a skeletal structure for "bent" butane based on this model. Answer the questions about these models as given on the report sheet.
3. Build a second model of butane in a straight chain. Remove one of the end carbon atoms from this model to give a chain of three carbon atoms. Remove a hydrogen atom from the central carbon of the three carbon chain. Connect the carbon you removed to the center carbon and the hydrogen to the end carbon of this three-carbon chain. Use your model to help you draw on your report sheet the skeletal structure of this branched chain model containing four carbons. Draw the condensed structure of this model on your report sheet.

4. Compare the model you built for "bent" butane with the model you built for the branched chain compound with four carbons. Answer the questions on the report sheet.

5. There are four structural isomers with the molecular formula C_4H_9Br. Build a model of each isomer, draw its skeletal structure, and give its correct IUPAC name in the space provided on the report sheet.

Part 3. Stereoisomers

A1. Build two models of 1,2-dimethylcyclopentane as follows: On the first model, add the methyls to the cyclopentane ring such that they are both on the same face (side) of the cyclopentane ring. On the second model, add the methyls to the cyclopentane ring such that they are on opposite faces (sides) of the cyclopentane ring.

A2. In the space provided on the report sheet, draw the skeletal structure of each compound using wedge-and-dash bonds to represent the position of the methyl groups. Answer the questions given on the report sheet.

B1. Build two identical four-carbon chains with the carbon atoms aligned in a straight line (similar to number 1 in Part 2 above). Place them side by side on the lab bench in front of you so that the same atoms on each model are touching the bench top.

B2. On the first model, attach a chlorine atom to the left side of its second carbon atom. On the second model, attach a chlorine atom to the right side of its second carbon atom. Complete both models by adding all of the hydrogens. In the space provided on the report sheet, write the molecular formula of the compound with the chlorine on the left.

B3. Using one of the models as your guide, draw a condensed structural formula of the compound. Answer the questions given on the report sheet.

B4. Using wedge-and-dash bonds on skeletal structures, in the space provided on your report sheet, draw a correct skeletal structure for each of the models.

Part 4. Shapes of Molecules

1. Build a model of each of the following molecules given the condensed structural formulas: CH_3NH_2 and CH_3OH.

2. Using the model you built as your guide, in the space provided on your report sheet, draw the Lewis structure for each molecule. Make sure you include all nonbonded electron pairs on your Lewis structures.

3. Compare the two models and answer the questions given on the report sheet.

Name _____

Date _____ Lab Section _____

Pre-Lab Questions | 5

1. Draw a Lewis structure for each of the molecules below. Give its VSEPR form and the geometry around the central atom.

   a. H_2O (water)

   b. NH_3 (ammonia)

2. Pentane has a molecular formula of C_5H_{12}. Answer the following questions about pentane.

   a. Draw the condensed structural formula for pentane.

   b. Draw a skeletal structure for pentane.

   c. Draw the condensed structural formula of an isomer of pentane that has a methyl branching from the main chain.

   d. Give the correct IUPAC name of the compound you drew in Part c.

   e. Does the molecule you drew in Part c contain a chiral carbon? If so, indicate the chiral carbon by placing * above it. If the molecule doesn't contain a chiral carbon, explain why.

Name _____

Date _____ Lab Section _____

REPORT SHEET | LAB

Molecular Models, Isomerism, and the Shape of Molecules

5

Part 1. Molecular Models of Simple Alkanes

Methane

Three-Dimensional Structure	Lewis Structure	Condensed Structural Formula

Give the VSEPR form of methane (using the carbon as "A"). _____

What is the geometry about the carbon atom in methane?_____

Ethane

Lewis Structure	Condensed Structural Formula

Propane

Lewis Structure	Condensed Structural Formula

What is the geometry about each of the carbon atoms in ethane?_____

What is the geometry about each of the carbon atoms in propane?_____

Using your model of propane, compare the C—C—C bond angle to one of the H—C—H bond angles. Are the angles the same, or are they different?

Part 2. Conformational and Structural Isomers

Skeletal Structure of Butane—Straight	Skeletal Structure of Butane—"Bent"

What type of isomers are the straight butane on the left and the "bent" butane on the right? How do you know?

Skeletal Structure of Branched Chain with Four Carbons	Condensed Structure of Branched Chain with Four Carbons

Give the correct IUPAC name of the branched chain compound. _____

Can you convert the branched chain model into the "bent" model without breaking a carbon–carbon bond?

What type of isomers are the branched chain compound and the bent compound? How can you be sure of your answer?

Draw the four structural isomers of C_4H_9Br and give the correct IUPAC name of each in the space provided.

Structure 1:	Structure 2:
Name:	Name:
Structure 3:	Structure 4:
Name:	Name:

Part 3. Stereoisomers

A2.

Skeletal Structures of 1,2-dimethylcyclopentane	

What type of isomers are represented by the two models?_____

Below the structures you drew above, give the correct IUPAC name of each of the compounds.

B2.

Molecular formula _____

Condensed Structural Formula

Do the two models represent molecules with the same molecular formula?

Is the connectivity of the atoms the same? In other words, does each carbon atom have the same type of atoms bonded to it in both models?

Are the two models of identical compounds?

If so, you should be able to superimpose one model on the other and see that all of the atoms on both models are in exactly the same arrangement in space. If not, how are the two models related to each other?

Skeletal Structures with Wedge-and-Dash Bonds	

Part 4. Shapes of Molecules

Lewis Structure of CH_3NH_2	Lewis Structure of CH_3OH

Give the VSEPR form of CH_3NH_2 (using the nitrogen as "A"). _____

Give the VSEPR form of CH_3OH (using the oxygen as "A"). _____

What is the geometry about the nitrogen atom in the first molecule?_____

What is the geometry about the oxygen atom in the second molecule?_____

Compare the C—N—H bond angle of the first molecule with the C—O—H bond angle of the second molecule? Are they identical? If not, which angle is smaller?

Compare the H—C—H bond angle of the methyl group (on either of the molecules) with the H—N—H bond angle of the CH_3NH_2. Are they identical? If not, which angle is smaller?

Compare the H—C—H bond angle of the methyl group (on either of the molecules) with the H—O—C bond angle of the CH_3OH. Are they identical? If not, which angle is smaller?

Based on your answers to the three previous questions, explain the effect of a nonbonded pair of electrons on bond angles and the shape of a molecule.

How Many Calories Are in Different Types of Nuts? | 6

LEARNING GOALS

After completing this laboratory exercise, you should be able to:

- Outline how a calorimeter is used to determine the heat energy released by a chemical reaction
- Explain how the Calorie content of food can be determined in the laboratory
- Discuss the relationship of a nutritional Calorie to a scientific calorie

BEFORE YOU BEGIN

In Chapter 1 of your text (*General, Organic, and Biological Chemistry, 2nd edition*), you were introduced to energy concepts from a chemistry point of view. We discussed the fact that foodstuffs contain energy that can be released to fuel our bodies. In Chapter 5, you saw that the energy in food can be measured by calorimetry, a technique in which an item of food is burned in a special apparatus that allows us to measure the amount of heat released. The heat released is directly related to the number of Calories contained in the food.

Before you begin this experiment, review the material from Chapters 1 and 5, focusing on the following sections and concepts:

- Section 1.4: Energy—kinetic and potential, the relationship of calories to dietary Calories, and specific heat
- Section 5.1: Heat of reaction, calorimetry, and energy content of food

INTRODUCTION

Energy is the capacity to do work. In chemistry, we often think of energy as the heat that is released or consumed by a chemical reaction. While not the only form of energy relevant to a chemist, heat energy is an important type of energy and is easy for most of us to relate to. For example, if you have used a chemical hand warmer on a cold day, you know that once you remove it from its foil packet, the hand warmer generates heat. The chemical reaction that occurs when the hand warmer is exposed to oxygen in the air releases heat into its surroundings. A reaction such as this that releases heat is known as an exothermic reaction. Similarly, a reaction that removes heat from its surroundings (gets colder as the reaction proceeds) is an endothermic reaction.

The energy released (or consumed) by a chemical reaction is measured in units of joules or in units of calories. In this experiment, we will use calories. A calorie is defined as the energy required to raise the temperature of 1 gram of water by 1 °C.

The energy produced by a chemical reaction can be measured using an instrument called a calorimeter. As shown in Figure 6.1, a calorimeter consists of an internal reaction chamber within a larger external container that is filled with water. When a chemical reaction occurs in the internal chamber of the calorimeter, the heat released by the reaction is absorbed by the water in the external container, raising the temperature of the water. The heat absorbed by the water is equal to the heat released by the reaction.

$$Q_{released} = Q_{absorbed} \qquad \text{Equation 1}$$

The change in the temperature of the water is related to the heat absorbed by the water by the equation

$$Q = mC\Delta T \qquad \text{Equation 2}$$

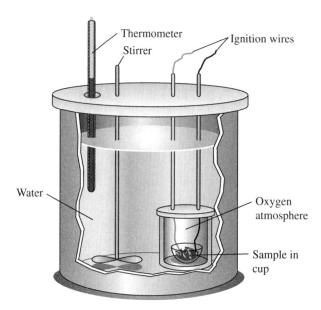

▲ **FIGURE 6.1** A calorimeter.

where Q represents heat energy (cal), m represents the mass (g) of the chemicals reacting, C represents a constant known as specific heat, and ΔT is the temperature change (°C). Knowing the mass of the water in the external container and measuring the change in the temperature of the water that occurs during the reaction, we can calculate the heat energy (calories) produced by the reaction.

Our bodies get the energy needed for the daily functions of life from the food we eat. This energy is obtained from food through the process of metabolism (see Chapter 12 of your text). The energy content of food is measured in kilocalories, also know as nutritional Calories, or simply Calories (note the capital C). A Calorie is equal to 1000 calories and is the amount of energy necessary to raise the temperature of 1 kilogram of water by 1 °C. In this experiment, we will construct a simple calorimeter and use it to measure the Calorie content of several types of nuts.

EXPERIMENT

Part 1. Preparing Your Calorimeter

1. Obtain an empty 12-ounce aluminum soft drink can. Rinse it thoroughly, dry the outside, and weigh it. Record the mass on your report sheet.
2. Using a graduated cylinder, measure 100 mL of water and pour it into the soft drink can.
3. Weigh the can + water and record the mass on your report sheet.
4. Using a three-finger clamp, clamp the soft drink can to a ring stand as shown in Figure 6.2.
5. Obtain a large paper clip and a cork stopper. Leaving the inside bend of the paper clip intact, straighten the outer bend of the paper clip so that it is perpendicular to the intact bend as shown in Figure 6.2.
6. Insert the straightened end of the paper clip into the small end of the cork stopper as shown in Figure 6.2 to produce a holder for your nut sample.
7. Place the holder directly beneath the can and adjust the height of the can on the ring stand so that the bottom of the can is about an inch above the holder.
8. Obtain a square piece of cardboard that will cover the top of the can; position it on top of the can; mark the cardboard above the opening in the can; and using a pencil or another sharp object, punch a small hole in the cardboard. The diameter of the hole should be large enough to accommodate a thermometer but small enough to hold the thermometer without allowing it to slip through the hole.
9. Insert a thermometer through the hole in the cardboard. (Take care when handling a glass thermometer. It is fragile and can easily break. Immediately report a broken thermometer to your instructor.) Place the cardboard back on top of the can and position the thermometer so that it is in the water but does not touch the bottom or sides of the can.

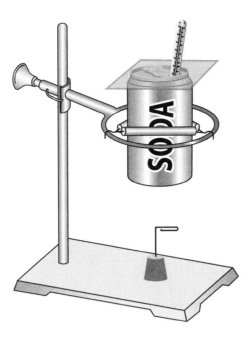

▲ **FIGURE 6.2** Soft drink can calorimeter.

Part 2. Measuring the Calorie Content of Nut Samples

1. On your report sheet, record the initial temperature of the water in the can.
2. Obtain a sample of the nut you will be testing. You will need a sample that is 0.5–1.5 g. (For small nuts such as peanuts and almonds, a single nut will be fine. For a large nut such as a walnut, cut the nut in half.)
3. Weigh the nut sample and record its mass on your report sheet.
4. Place the nut sample on the holder, making sure it is stable and will not easily fall off the holder.
5. Using a wooden kitchen match or a propane lighter, ignite the nut sample. This may take several seconds. Be patient and keep the flame on the nut sample until it ignites.
6. Allow the nut sample to burn completely. (Reignite the nut sample if the flame goes out before the nut sample is completely burned.)
7. Immediately after the nut sample stops burning, use the thermometer to carefully stir the water in the can. Determine the temperature of the water at its highest temp and record this value on your report sheet as the final temperature of water.
8. Allow the charcoal-like residue from the nut sample to cool to room temperature. Determine the mass of the residue and record its mass on your report sheet.
9. Allow the can to cool to room temperature, remove the cardboard and thermometer, clean the black soot off the can, and pour the water from the can down the drain.
10. Repeat Parts 1 and 2 for your second nut sample.

Part 3. Calculations

Show calculations in the space provided after the data page on your report sheet.

1. Calculate the mass of water used in each trial by subtracting the mass of the can from the mass of the can + water (*Part 1.C – Part 1.B*). Record the mass of the water on your report sheet.
2. Calculate the temperature change of the water in each trial by subtracting the initial temperature of the water from the final temperature of the water (*Part 2.C– Part 2.A*). Record this value on your report sheet.
3. Using Equation 1, calculate the heat energy (Q) absorbed by the water in the calorimeter. On your report sheet, record the energy absorbed by water. (For this calculation, m is the mass of the water and ΔT is the temperature change of the water. The value of C is given on the report sheet.)

4. Record the energy released by nut sample on your report sheet. (Keep in mind that the energy released by the burning nut sample is approximately equal to the energy absorbed by the water: $Q_{released} = Q_{absorbed}$.)

5. Convert the energy released to nutritional Calories. (Use this conversion factor: 1 Calorie/1000 calories.)

6. Calculate the mass of the nut sample consumed by combustion by subtracting the mass of the residue after the nut is burned from the mass of the nut (*Part 2.B– Part 2.D*). On your report sheet, record the mass of nut consumed by combustion.

7. Calculate the Calorie content of the nut sample by dividing the energy released by nut sample (Cal) by the mass of nut consumed by combustion (g) (*Part 3.F/Part 3.G*). Record this value on your report sheet.

Pre-Lab Questions | 6

1. Explain the difference between an endothermic and an exothermic reaction.

2. A student calculated that the water in a calorimeter absorbed 8 kcal of heat energy when a sample of potato chips was burned in the calorimeter. How many kcal of heat energy did the potato chips release when burned?

3. How many calories are in a sample of a snack food containing 125 Calories?

4. What nutritional component in nuts do you suppose allows them to burn easily when ignited with a match or propane lighter?

Name _____

Date _____ Lab Section _____

REPORT SHEET | LAB

How Many Calories Are in Different Types of Nuts? | 6

Part 1. Preparing Your Calorimeter

	Trial 1	Trial 2
A. Type of nut used	_____	_____
B. Mass of empty soft drink can	_____	_____
C. Mass of can + water	_____	_____

Part 2. Measuring the Calorie Content of Nut Samples

	Trial 1	Trial 2
A. Initial temperature of water, °C	_____	_____
B. Final temperature of water, °C	_____	_____
C. Mass of nut, g	_____	_____
D. Mass of residue after nut is burned, g	_____	_____

Part 3. Calculations

	Trial 1	Trial 2
A. Mass of water, g	_____	_____
B. Temperature change (ΔT), °C	_____	_____
C. Specific heat of water	1.0 calorie/gram °C	
D. Energy absorbed by water, cal	_____	_____
E. Energy released by nut sample, cal	_____	_____
F. Energy released by nut sample, Cal (nutritional)	_____	_____
G. Mass of nut consumed by combustion, g	_____	_____
H. Calorie content of nut sample (Cal/g)	_____	_____

Show your calculations for Part 3 in the space below.

QUESTIONS

1. Find two other people in your lab section who used each type of nut that you used. Compare your value for the Calorie content of each of the different nuts with theirs. Are the values similar? If not, why? Calculate an average value for the Calorie content of each type of nut.

2. Find the actual Calorie content of the types of nuts you used by checking the label on the container or by accessing the website of the National Nutrient Database (http://ndb.nal.usda.gov) or a website such as CalorieKing (http://www.calorieking.com). Using your value and the average value you calculated in question 1, compare the actual value to your experimental value. Explain any differences in the values.

3. Some of the heat generated from the burning of your nut sample may have been lost to the air or the can.

   a. How would this loss of heat affect the value of the Calorie content you calculated?

   b. How might you have prevented this heat loss?

How Are Condensation Reactions Used Beneficially? | 7

LEARNING GOALS

After completing this laboratory exercise, you should be able to:

- Draw the structures of aspirin and methyl salicylate and identify the functional groups of each
- Explain how to calculate a percent yield and why it is important
- Describe how to take a melting point and explain why it is useful
- Explain how chemists use condensation reactions

BEFORE YOU BEGIN

In Chapter 4 of your text (*General, Organic, and Biological Chemistry, 2nd edition*), you were introduced to the variety of families of organic compounds and some of their uses, including their role as pharmaceuticals. In Chapter 5, you encountered some reactions of organic compounds, including condensation reactions.

Before you begin this experiment, review the material from Chapters 4 and 5, focusing on the following sections and concepts:

- Section 4.3: Families of organic compounds and organic functional groups, especially alcohols, phenols, carboxylic acids, and esters
- Section 5.5: Reactions of organic compounds, particularly condensation reactions

INTRODUCTION

The majority of medically important compounds, such as the over-the-counter (OTC) medicines you buy at the pharmacy as well as the medicines your doctor prescribes, are organic compounds. Reading the inserts in prescription medications, you will discover that most pharmaceuticals are structurally complex organic molecules containing several functional groups.

Medically important compounds such as the chemotherapeutic agent Taxol® (generic paclitaxel), which is used in the treatment of breast cancer, can often be obtained from plant sources. In the case of Taxol®, the drug was originally isolated from the bark of the Pacific Yew tree in a process that destroyed the tree. The tremendous potential of Taxol® in treating cancer coupled with the scarcity of the source from which it was isolated led scientists to begin searching for other ways to obtain the drug. The strategy most often employed by organic chemists is synthesis—making the naturally occurring compound in the laboratory from readily available starting materials.

Synthesis of a complex organic molecule often involves one or more condensation reactions. In a condensation reaction, two (or more) smaller molecules are combined to create the larger, more complex molecule often forming water (or other small molecules) as a by-product. In this experiment, you will synthesize two familiar over-the-counter medicines from the same starting compound, salicylic acid. As shown in Scheme 7.1, each of the syntheses involves a condensation reaction of two molecules to form the desired compound.

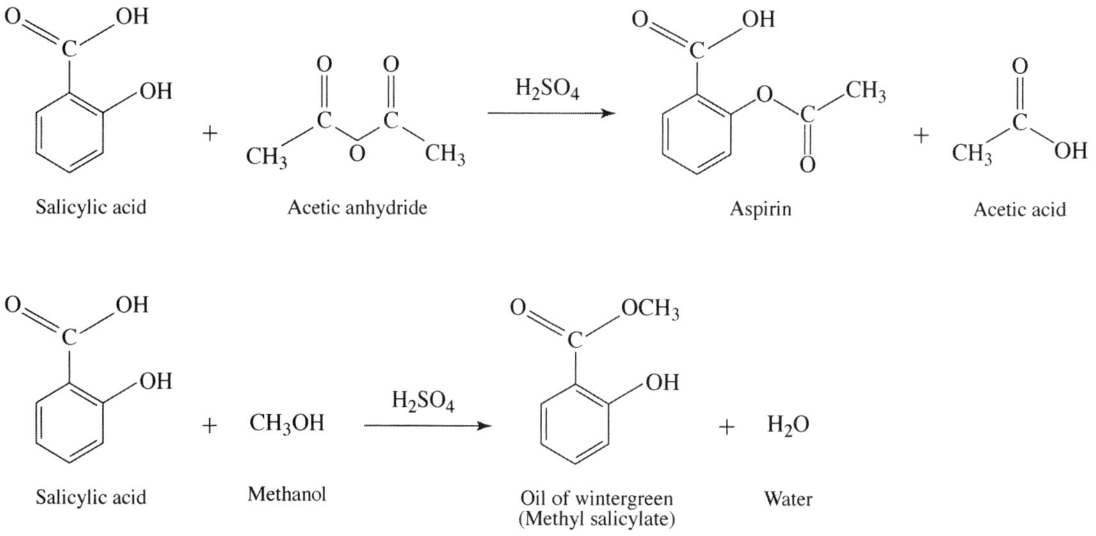

▲ **SCHEME 7.1** Condensation reactions used in the syntheses of aspirin and oil of wintergreen.

In Part 1 of the lab, you will synthesize aspirin, the well-known over-the-counter medicine, from salicylic acid and acetic anhydride. Aspirin is widely used to relieve pain and reduce inflammation. It is also effective in reducing fevers in adults, but is not used for this purpose with children due to the risk of developing Reye's syndrome, a potentially fatal disease that affects the brain and liver.

In Part 2 of the lab, you will synthesize methyl salicylate, also known as oil of wintergreen. Methyl salicylate is a versatile compound that is used medically as a rubefacient in ointments such as Icy Hot® and Bengay. Methyl salicylate is also used as a flavoring agent in chewing gum, candy, and root beer; as an insect repellent; and as an antiseptic in mouthwash. It is incorporated into botanical perfumes as well.

When synthesizing a compound in the lab, chemists pay particular attention to two important factors— the purity of the product and the percent yield. To ensure that the product produced in the lab is the desired compound, chemists use a variety of methods to confirm the identity of the synthetic compound and its purity. In today's lab, you will test the melting point of the aspirin you synthesize. Melting point is a characteristic property of a compound and, therefore, provides a straightforward method for confirming the identity of the compound. In addition, because impurities in a compound will lower its melting point and/or broaden the range of temperatures over which a compound melts, melting point is also an effective method for determining the purity of a compound.

To maximize resources and potential profitability, chemists monitor the percent yield of a synthesis. Percent yield is simply a ratio of the grams of product obtained from the reaction (actual yield) and the maximum possible grams of product obtainable from the reaction (theoretical yield) converted to a percentage, as shown in Equation 1. A percent yield over 80% is generally considered a good yield.

$$\frac{\text{actual yield (g)}}{\text{theoretical yield (g)}} \times 100 = \text{percent yield} \qquad \text{Equation 1}$$

EXPERIMENT

Part 1. Synthesis of Aspirin

1. Set up a hot water bath as follows: Fill a 400 mL beaker half full of water. Place the beaker on a hot plate and heat until the water is simmering. (Do not heat the water to a full boil, just a gentle boil.) Keep the water at this gentle boil, taking care that the beaker does not boil dry, until you complete all parts of this experiment.

2. Set up an ice water bath as follows: Fill a 400 mL beaker half full of ice and water. Fill a 250 mL beaker half full of distilled water and immerse the beaker in the ice bath to chill the distilled water.

3. Obtain a clean, dry 125 mL Erlenmeyer flask. Label the flask with your name.

4. Weigh out approximately 2 g of salicylic acid. Record the mass to the nearest 0.01 g on your report sheet.

5. Transfer the salicylic acid into the Erlenmeyer flask and take the flask and a clean, dry 10 mL graduated cylinder to the fume hood.

6. Working in the hood with gloves on, measure 5 mL of acetic anhydride with the graduated cylinder and pour it slowly down the inside of the Erlenmeyer flask containing the salicylic acid. Swirl the flask to mix the contents. (CAUTION: Acetic anhydride is corrosive and an irritant, especially to the nose and eyes. Do not breathe the vapors. Handle with care.)

7. Still working in the hood, add 5 or 6 drops of concentrated H_2SO_4 (sulfuric acid) to the flask and swirl gently to mix. (CAUTION: Sulfuric acid is highly corrosive! Wear gloves and handle with care.)

8. Cover the mouth of the Erlenmeyer flask containing the salicylic acid, acetic anhydride, and sulfuric acid with a small watch glass, hold the watch glass in place with your finger, and place the flask into the hot water bath. Swirl until all of the solid dissolves.

9. Continue to heat and swirl the mixture for 5 minutes after all of the solid has dissolved.

10. Remove the flask from the hot water bath and return it to the fume hood. Allow it to cool until it is no longer hot to the touch.

11. Lower the heat on the water bath. (For Part 2, the water bath needs to be 70°C.)

12. Working in the hood, add 50 mL of the chilled distilled water to the Erlenmeyer flask containing the reaction mixture. Stir with a glass stirring rod. You may note white crystals beginning to form.

13. Place the flask and its contents into the ice bath until you no longer see crystals forming. (This may take 5–10 minutes.) If crystals do not form, scratch the inside of the flask with a glass stirring rod.

14. While you are waiting for the crystals to form, set up a vacuum filtration apparatus as shown in Figure 7.1.

15. Place a piece of filter paper on the funnel and turn on the vacuum source. Dampen the paper slightly to help it seal to the funnel.

16. Slowly pour the contents of the Erlenmeyer flask onto the filter paper. The crystals of aspirin will collect on the paper. Use chilled distilled water to rinse the flask and transfer all of the solid to the filter paper.

17. Once all of the solid has been transferred onto the filter paper, rinse it 2 or 3 times by slowly pouring 5 mL of the chilled distilled water over the solid. The crystals should be spread evenly over the filter paper at this point. If not, use your stirring rod to gently spread them out. Be careful not to tear the filter paper.

18. Leave the vacuum source turned on to draw air over the solid to help dry it while you perform the next part of the experiment. Crystals may be dried in a warm oven if available.

19. After the solid crystals are completely dry, weigh your dried aspirin and record the mass on your report sheet.

20. Describe the crystals in the indicated space on your report sheet.

21. Obtain a melting point for the aspirin that you synthesized. (Your instructor will demonstrate the appropriate procedure based on the equipment in your lab.) Pure aspirin melts at 135°C. Salicylic acid melts at 159°C. (Impurities lower the melting point of a substance even if the impurity has a higher melting point.) Record the melting point range on your report sheet.

22. Dispose of your aspirin and all other waste from this experiment as directed by your instructor.

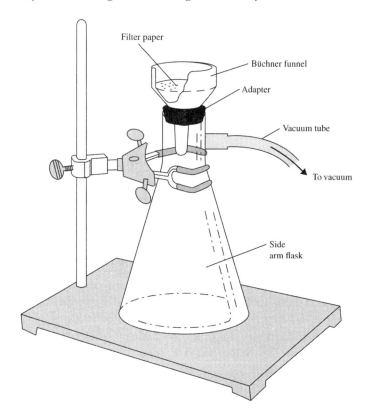

▲ **FIGURE 7.1** A vacuum filtration setup.

Part 2. Synthesis of Oil of Wintergreen (Methyl Salicylate)

1. Weigh out approximately 0.5 g of salicylic acid and place it in a large, clean, dry test tube.
2. Carefully smell the contents of the tube using the technique you learned in Lab 1. Note any characteristic odor and record your observations on your report sheet.
3. Add 2 mL of methanol to the test tube. Stopper the tube with a cork stopper and mix the contents by holding the top of the tube firmly and tapping the bottom with your finger. Mix until the salicylic acid completely dissolves.
4. Again, smell the contents of the tube using the technique you learned in Lab 1. Note any characteristic odor and record your observations on your report sheet.
5. Working in the hood with gloves on, carefully add 3 or 4 drops of H_2SO_4 (sulfuric acid) to the tube, stopper the tube with a cork stopper, and mix the contents of the tube as you did in step 3 immediately above. (CAUTION: Sulfuric acid is highly corrosive! Wear gloves and handle with care.)
6. Place the test tube in the warm water bath, which should be at 70°C, for 15 minutes. Point the mouth of the test tube away from others and observe the tube and contents to ensure that nothing boils over.
7. Remove the tube from the water bath and allow it to cool to room temperature. Turn off the heat on your water bath.
8. Once the tube has cooled, smell the contents of the tube using the technique you learned in Lab 1. Note any characteristic odor and record your observations on your report sheet.
9. Dispose of the contents of the test tube as directed by your instructor.

Part 3. Calculations

1. Calculate your theoretical yield of aspirin using the following formula:

$$(\text{grams of salicylic acid used}) \times \frac{180}{138} = \text{theoretical yield (g)}$$

2. Calculate your percent yield of aspirin using the formula in the Introduction.
3. Record your percent yield on your report sheet.

Pre-Lab Questions | 7

1. Circle and give the name of each functional group in the aspirin molecule shown below.

Aspirin

2. Draw the structure of benzoic acid and phenol in the space below. Draw the structure of salicylic acid next to the other two. Using your drawings, explain the structural similarities between salicylic acid and each of the other molecules.

3. Give the product and by-product of the following condensation reaction. Refer to your text if needed.

$+$ CH_3OH $\xrightarrow{H_2SO_4}$

4. List two factors that are important to a chemist when synthesizing a compound in the lab. Describe why each is important.

REPORT SHEET | LAB

How Are Condensation Reactions Used Beneficially?

7

Part 1. Synthesis of Aspirin

Mass of salicylic acid   _____ g

Mass of dried aspirin (product)   _____ g

Percent yield   _____ %

Melting point range of your dried aspirin   _____ °C

Show the calculation of your theoretical and percent yields in the space below.

Describe your final aspirin sample.

Part 2. Synthesis of Oil of Wintergreen (Methyl Salicylate)

Observations of salicylic acid (smell, appearance, etc.)

Observations of salicylic acid + methanol (before addition of H_2SO_4)

Observations of reaction mixture (salicylic acid, methanol, and H_2SO_4) after heating and cooling back to room temperature

QUESTIONS

1. What does the melting point range of your aspirin tell you about the purity of your product? What are some potential impurities in your final product? (Hint: Consider the compounds used in the procedure.)

2. Draw the structures of aspirin and methyl salicylate in the space below. Describe how the structures of the two compounds are similar and how they are different.

3. Describe the properties of aspirin and methyl salicylate (state of matter, smell, and appearance). Would you describe them as similar or different? From what you know about their uses, do they have similar modes of action as medicines?

4. Sulfuric acid was used in both of the reactions you performed today, but it was used in a very small amount. What purpose did sulfuric acid serve in the reactions? (Hint: Consider where sulfuric acid is written in the reaction equation.)

What Happens to Bananas as They Ripen? | 8A

LEARNING GOALS

After completing this laboratory exercise, you should be able to:

- Explain the differences between the chemical composition of the flesh of an unripe banana and a ripe banana
- Name the type of chemical reaction that occurs during the ripening of a banana
- Identify the chemical bond that is broken during the ripening of a banana

BEFORE YOU BEGIN

In Chapter 6 of your text (*General, Organic, and Biological Chemistry, 2nd edition*), you were introduced to the structures and reactions of different types of carbohydrates. Before you begin this experiment, review the material in Chapter 6 focusing on the following sections and concepts:

- Section 6.1: the similarities and differences between the structures of monosaccharides and polysaccharides
- Section 6.5: identification of a glycosidic bond; name of the type of reaction that forms a glycosidic bond and the type of reaction that breaks a glycosidic bond
- Section 6.6: the structure of starch and the changes in that structure during ripening of fruit; name of the enzyme responsible for changes in starch during the ripening process

INTRODUCTION

If you are a banana eater, you know that the flavor of a green banana is much different from the flavor of a yellow or brown banana. You also know that if you buy bananas well ahead of when you plan to eat them, they may well turn brown and mushy before you get the chance to eat them. The processes that change the flavor of the banana and the color of the banana's skin are chemical processes. In this lab, we will investigate these changes and learn what is occurring chemically in the banana as it ripens.

In this laboratory, each group will be assigned a type of banana [green (unripe), yellow (ripe), or brown (overripe)] to test for starch and glucose content. At the end of the lab, you will average your results and share your findings with the rest of the class. In this part of the project, you will discover what happens to the chemical composition of a banana as it ripens.

To measure the glucose content of the prepared samples, you will use commercially available test strips that are typically used by diabetic patients to monitor glucose levels in their urine. These test strips are impregnated with an enzymatic system that reacts with glucose and produces different colors based on the concentration of glucose present. The concentration of glucose is determined by comparing the color of the test strip with the color chart on the bottle containing the test strips.

EXPERIMENT

Step 1: Isolation of soluble sugars from bananas

1. Obtain an unripe (green), ripe (yellow), or overripe (brown) banana, as assigned, from your instructor.

2. Determine the mass of approximately 1 g of banana on a top-loading balance. ***Note: It is not necessary to weigh out exactly 1 g of the banana; however, it is important to know the exact mass of your sample.***

3. Record the mass of the banana sample to two decimal places on the report sheet.

4. Place this banana sample in a mortar, add 6 mL of distilled water, and thoroughly grind the sample using the pestle.

5. Pour the ground-up sample from the mortar into a centrifuge tube (or test tube).

6. Rinse the mortar with 6 mL of fresh distilled water and add the rinse to the same centrifuge tube or test tube.

7. Centrifuge the sample in the tube (make sure the centrifuge is balanced as demonstrated by your instructor) for 10 minutes at room temperature.

8. Pour off the supernatant (the liquid on top) into a clean 125 mL Erlenmeyer flask, leaving any solid (the pellet).

9. Wash the pellet (the solid left behind) with 6 mL of fresh distilled water by adding the water to the centrifuge tube or test tube and thoroughly breaking up the pellet with a glass stirring rod. Be careful not to break the centrifuge tube or the stirring rod! Rinse the stirring rod with a small amount of water, allowing the water to fall into the centrifuge tube.

10. Centrifuge the sample again for 10 minutes.

11. Pour off this supernatant, adding it to the one from the previous centrifugation (in the 125 mL Erlenmeyer flask).

12. In a safe place, set aside the pellet in the centrifuge tube for later use.

13. Using a graduated cylinder, carefully measure the volume of the combined supernatants (in the 125 mL Erlenmeyer flask).

14. Accurately record this volume on the report sheet.

15. Pour the solution back into the 125 mL Erlenmeyer flask, label it "Banana extract," and set it aside in a safe place for later use.

Step 2: Isolation of the insoluble starch

Nearly all of the insoluble matter in bananas is composed of starch. During this step, your task is to isolate the starch.

1. Resuspend the pellet in the centrifuge tube in 10 mL of distilled water using a glass stirring rod (or a Vortex-Genie as demonstrated by your instructor). Be careful not to break the centrifuge tube or the stirring rod!

2. Obtain a piece of filter paper, use a pencil to write your name on it, determine the mass of the paper, and record this mass on the report sheet.

3. Place the filter paper on top of a Büchner funnel on a vacuum filtration apparatus (see Figure 8.1) and filter the resuspended pellet.

4. Add 10 mL of distilled water to the centrifuge tube; rinse; and pour this through the funnel, taking care to transfer as much of the solid as possible.

5. Wash the solid on the filter paper with a stream of acetone from a squeeze bottle to help aid in drying.

6. If necessary, place the sample in an oven to speed up the drying process. DO NOT determine the mass until the filter paper is completely dry.

7. When the filter paper is completely dry, determine the mass of the paper with the dried solid and accurately record the mass to two decimal places on your report sheet.

Step 3: Measuring the concentration of glucose in the banana extract

Most of the soluble material in your sample is glucose. Depending on the type of banana you test, you may need to dilute your sample. You will be using Keto-Diastix Reagent® Strips to test for glucose.

1. Remove one test strip from the bottle. DO NOT touch the test areas of the strip.

2. Dip the test strip in the banana extract and remove immediately (drawing the edge of the strip against the rim of the beaker or flask).

3. Compare the glucose test area on the strip to the color chart (on the bottle) at the time directed on the bottle. (Typically, the test strip is read **exactly 30 seconds** after wetting.)

4. Record the glucose concentration on your report sheet.

5. If the concentration of glucose exceeds what the test strips can accurately measure, you will need to dilute your sample and retest the glucose concentration. See your instructor to receive instructions about sample dilution.

6. Dispose of all waste including solutions and test strips as directed by your instructor.

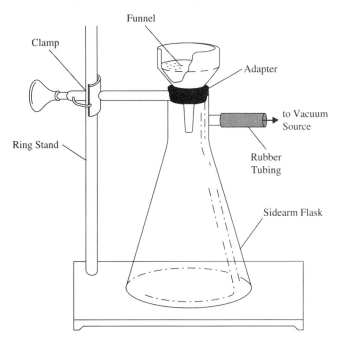

▲ **FIGURE 8.1** Vacuum filtration apparatus

Pre-Lab Questions | 8A

1. Explain the structural relationship between monosaccharides and polysaccharides.

2. When two monosaccharides are chemically joined together to form a disaccharide, what type of bond is formed between the two monosaccharides? Draw the structure of a disaccharide that could be formed from two glucose molecules and circle the bond that was formed.

3. Name the type of reaction that occurs when two glucose molecules react to form a disaccharide (as in the preceding question).

4. Name the type of reaction that occurs when a polysaccharide is broken down into monosaccharides and disaccharides.

Name _____

Date _____ Lab Section _____

REPORT SHEET | LAB

What Happens to Bananas as They Ripen?

| 8A

Part 1

Step 1

Banana Type (Green, Ripe, Overripe)   _____

Color of Banana Skin   _____

Mass of Banana Sample   _____ g

Volume of Banana Extract (in mL)   _____

Step 2

Mass of Filter Paper   _____ g

Mass of Starch and Filter Paper   _____ g

Mass of Starch   _____ g

% Starch in Banana (g starch/g of banana sample) × 100   _____ %

Step 3

Concentration of Glucose in Banana Extract   _____ mg/dL

Total Grams of Glucose in Banana Extract*   _____ g

    (Conc. of Glucose × Volume of Banana Extract)

% Glucose in Banana (g glucose/g of banana sample) × 100   _____ %

*Prior to beginning this calculation, convert the Conc. of Glucose from mg/dL (as given by the test strips) to g/mL.

Class Data

Type of Banana	Avg. % Glucose	Avg. % Starch
_____	_____	_____
_____	_____	_____
_____	_____	_____

Sample calculations

Clearly show the calculations you made to determine the "Total grams of glucose in banana extract" and the "% glucose in banana."

QUESTIONS

1. Based on your experimental results, what happens to the chemical composition of a banana as it ripens?

2. What type of chemical reaction is taking place (inside the banana) as it ripens?

3. What type of bond is cleaved during this process?

4. Using what you have learned in this experiment, explain why a ripe and/or overripe banana tastes sweeter than a green banana.

Can We Control the Ripening of Bananas? | 8B

LEARNING GOALS

After completing this laboratory exercise, you should be able to:

- Identify which, if any, storage method can be used to slow the ripening of a banana
- Evaluate the overall effectiveness of each storage method for maintaining the banana's chemical composition (which is directly related to its taste), appearance, and desirability for consumption

BEFORE YOU BEGIN

In Chapter 6 of your text (*General, Organic, and Biological Chemistry, 2nd edition*), you were introduced to the structures and reactions of different types of carbohydrates. Before you begin this experiment, review the material in Chapter 6 focusing on the following sections and concepts:

- Section 6.6 – the structure of starch and the changes in that structure during ripening of fruit; name of the enzyme responsible for changes in starch during the ripening process

INTRODUCTION

In the previous experiment, you discovered the change in the chemical composition of the flesh of a banana that occurs upon ripening. In this experiment, you will investigate whether it is possible to control the ripening process.

Each group will start this experiment by selecting a green (unripe) banana and storing that banana for one week. Your group will be assigned to store its banana in the refrigerator; on a banana stand; in a cool, dark drawer; or in a closed brown paper bag. At the end of one week, you will test the banana for starch and glucose content using the procedures from the previous experiment. Once again, you will average your results and share them with your classmates. In this part of the project, you will use the knowledge you gained in the first part of the project and apply it to determine the best way to store a banana to keep it from ripening too quickly and to preserve its typical yellow appearance.

EXPERIMENT

Step 1: Banana storage

1. Obtain a green (unripe) banana from a single bunch. All of the bananas should be as equally ripe as possible, as determined by visual inspection.
2. Store your banana using one of the storage methods detailed by your instructor: refrigerator; banana stand; cool, dark drawer; or closed brown paper bag. Place your banana in storage and label it carefully so that others who use the lab will not disturb it.

Step 2: Isolation of soluble sugars from bananas

1. After your banana has been in storage for one week, carefully remove it from storage and open the peel.
2. Determine the mass of approximately 1 g of banana on a top-loading balance. ***Note: It is not necessary to obtain exactly 1 g of the banana; however, it is important to know the exact mass of your sample.***
3. Record the mass of the banana sample to two decimal places on your report sheet.
4. Place this banana sample in a mortar, add 6 mL of distilled water, and thoroughly grind the sample using the pestle.
5. Pour the ground-up sample from the mortar into a centrifuge tube (or test tube).

6. Rinse the mortar with 6 mL of fresh distilled water and add the rinse to the same centrifuge tube or test tube.

7. Centrifuge the sample in the tube (make sure the centrifuge is balanced) for 10 minutes at room temperature as demonstrated by your instructor.

8. Pour off the supernatant (the liquid on top) into a clean 125 mL Erlenmeyer flask, leaving behind any solid (the pellet).

9. Wash the pellet (the solid left behind) with 6 mL of fresh distilled water by adding the water to the centrifuge tube or test tube and thoroughly breaking up the pellet with a glass stirring rod. Be careful not to break the centrifuge tube or the stirring rod!

10. Centrifuge the sample again for 10 minutes.

11. Pour off this supernatant, adding it to the one from the previous centrifugation (in the 125 mL Erlenmeyer flask).

12. In a safe place, set aside the pellet in the centrifuge tube for later use.

13. Using a graduated cylinder, carefully measure the volume of the combined supernatants (in the 125 mL Erlenmeyer flask).

14. Accurately record this volume on your report sheet.

15. Pour the solution back into the 125 mL Erlenmeyer flask, label it "Banana extract," and set it aside in a safe place for later use.

Step 3: Isolation of the insoluble starch

Nearly all of the insoluble matter in bananas is composed of starch. During this step, your task is to isolate the starch.

1. Resuspend the pellet in the centrifuge tube in 10 mL of distilled water using a glass stirring rod (or a Vortex-Genie) as demonstrated by your instructor. Be careful not to break the centrifuge tube or the stirring rod!

2. Obtain a piece of filter paper, use a pencil to write your name on it, determine the mass of the paper, and record this mass on your report sheet.

3. Using the same filtration apparatus you used in the previous experiment (see Figure 8.1), place the filter paper on the filtration setup and filter the resuspended pellet.

4. Add 10 mL of distilled water to the centrifuge tube; rinse; and pour this through the funnel, taking care to transfer as much of the solid as possible.

5. Wash the solid on the filter paper with a stream of acetone from a squeeze bottle to help aid in drying.

6. If necessary, place the sample in an oven to speed up the drying process. DO NOT determine the mass until the filter paper is completely dry.

7. When the filter paper is completely dry, determine the mass of the paper with the dried solid and accurately record the mass to two decimal places on your report sheet.

Step 4: Measuring the concentration of glucose in the banana extract

Most of the soluble material in your sample is glucose. Depending on the type of banana you test, you may need to dilute your sample. You will be using Keto-Diastix® Reagent Strips to test for glucose.

1. Remove one test strip from the bottle. DO NOT touch the test areas of the strip.

2. Dip the test strip in the banana extract and remove immediately (drawing the edge of the strip against the rim of the beaker or flask).

3. Compare the glucose test area on the strip to the color chart (on the bottle) at the time directed on the bottle. (Typically, the test strip is read **exactly 30 seconds** after wetting.)

4. Record the glucose concentration on the data sheet.

5. If the concentration of glucose exceeds what the test strips can accurately measure, you will need to dilute your sample and retest the glucose concentration. See your instructor to receive instructions about sample dilution.

6. Dispose of all waste including solutions and test strips as directed by your instructor.

REPORT SHEET | LAB

Can We Control the Ripening of Bananas? | 8B

Method of storage (refrigerator, stand, drawer, paper bag)   _____

Step 2

Mass of Banana Sample   _____ g

Volume of Banana Extract (in mL)   _____ g

Appearance of banana after storage   _____

Step 3

Mass of Filter Paper   _____ g

Mass of Starch and Filter Paper   _____ g

Mass of Starch   _____ g

% Starch in Banana (g starch/g of banana sample) × 100   _____ %

Step 4

Concentration of Glucose in Banana Extract   _____ mg/dL

Total Grams of Glucose in Banana Extract*   _____ g

    (Conc. of Glucose × Volume of Banana Extract)

% Glucose in Banana (g glucose/g of banana sample) × 100   _____ %

*Prior to beginning this calculation, convert the Conc. of Glucose from mg/dL (as given by the test strips) to g/mL.

Class averages for the percentages of glucose and starch in stored bananas

Percentage of Glucose and Starch in a Banana

Carbohydrate	Refrigerator	Stand	Drawer	Paper Bag
Glucose	_____	_____	_____	_____
Starch	_____	_____	_____	_____

QUESTIONS

1. Based on the change in the chemical composition of the banana, what is the best method for storing a banana? Explain your answer.

2. Why did the method you identified in question 1 above slow the ripening of the banana? (For a hint, refer to section 6.6 in your text and read about what causes the ripening of fruit.)

3. Did the appearance of the banana after storage vary among the storage methods? If so, which method best preserved the appearance of the banana?

4. Based on your answers to the questions above, if you were writing for a national food magazine, how would you recommend storing bananas? When answering this question, consider the factors that are important to consumers (taste, appearance, etc.).

What Are the Best Stain Removers? | 9

LEARNING GOALS

After completing this laboratory exercise, you should be able to:

- State and explain the golden rule of solubility
- Predict the polarity of the molecules composing a stain based on the polarity of the solvent used to remove the stain

BEFORE YOU BEGIN

In Chapter 3 of your text (*General, Organic, and Biological Chemistry, 2nd edition*), you discovered that the polarity of a *molecule* is determined by the distribution of electrons within the molecule. You also saw that the polarity of a molecule can be predicted on the basis of its shape and the polarity of its bonds. In Chapter 7, you were introduced to several types of attractive forces and learned how those forces affect the physical properties of a compound, especially solubility. In this lab, you will apply these principles to investigate the golden rule of solubility and its role in removing stains.

Before you begin this experiment, review the material from Chapters 3 and 7, focusing on the following sections and concepts:

- Section 3.7: How to determine whether a molecule is polar or nonpolar
- Section 7.1: Types of attractive forces and how to identify the attractive forces present in a molecule
- Section 7.3: The golden rule of solubility and how to predict the solubility of a compound

INTRODUCTION

Attractive forces between molecules are the essential forces that hold our DNA in its double helical structure and keep our cell membranes intact. The physical properties of a substance, such as its boiling point, also depend largely on the attractive forces between its molecules. In turn, the attractive forces present in a molecule derive from the polarity of the molecule, which we learned in Chapter 3 is determined by the distribution of electrons in the molecule. A molecule with an even distribution of electrons over all of its atoms is nonpolar and has weak attractive forces. A molecule with an uneven distribution of electrons over its atoms is polar and has stronger attractive forces. In this lab, we will investigate the effect that attractive forces have on the solubility of a substance in a variety of solvents.

The golden rule of solubility—*like dissolves like*—is a simple way of saying that substances with similar polarities will dissolve in each other. In other words, a substance that is polar will dissolve in a solvent that is also polar but will not dissolve in a solvent that is nonpolar. Likewise, a nonpolar substance will dissolve in a nonpolar solvent but not in a polar one. The key to understanding the golden rule and solubility is understanding the attractive forces that exist between molecules.

Molecules that have similar attractive forces will readily interact with each other. In Chapter 7, we learned that table sugar (sucrose) dissolves in water but not oil because both sucrose and water are polar molecules that can form hydrogen bonds (see Figure 9.1 which depicts some of the possible hydrogen bonds). Because the water and sucrose can interact by forming multiple hydrogen bonds with each other, sucrose will dissolve in water. However, oil molecules are largely nonpolar hydrocarbons, and their only attractive force is London forces. Because oil and sucrose do not have similar attractive forces, they cannot readily interact with each other. This lack of interaction means that sucrose is not soluble in oil.

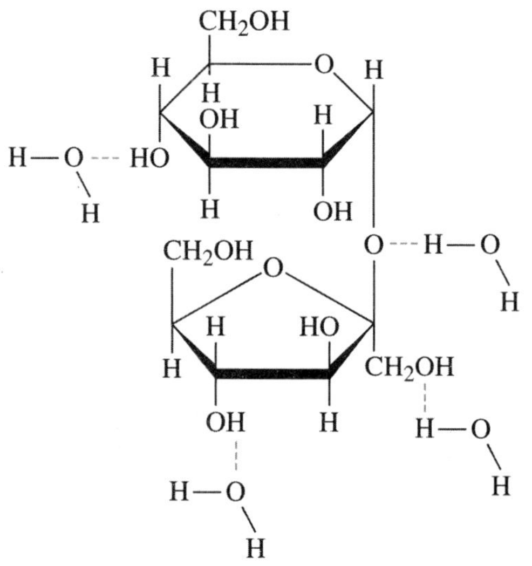

▲ **FIGURE 9.1** Sucrose can form multiple hydrogen bonds with water. A few of the many possible hydrogen bonds are shown (with the dashed-lines) above.

The polarity of any molecule falls at some point along a continuum between polar and nonpolar. As such, stains in your clothing are composed of molecules that also fit on this continuum. Removing a stain from your clothing is simply a matter of knowing whether the molecules of the stain are polar or nonpolar and then selecting the proper solvent to dissolve those molecules while leaving your clothing (and its coloring!) intact.

In this lab, you will investigate four common stains—ink from a ballpoint pen, chocolate syrup, cooking oil, and lipstick—and their solubility in some household solvents (see Figure 9.2), including hexane, a nonpolar hydrocarbon solvent often found in dry cleaning fluid; ethyl acetate, a moderately polar organic compound found in many nail polish removers; and isopropyl alcohol, a polar organic compound found in rubbing alcohol. (For comparison, you will also test the effectiveness of soapy water at removing the stains.) Because you know the polarity of the solvents you will be using, you should be able to infer the polarity of the molecules composing each stain by determining which solvent is most effective at removing it.

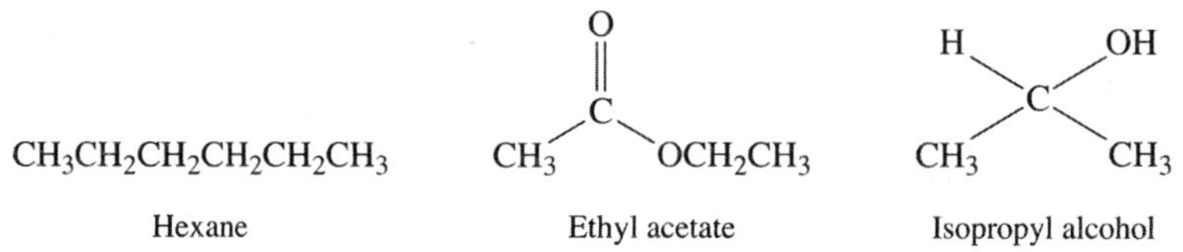

$CH_3CH_2CH_2CH_2CH_2CH_3$ Hexane   Ethyl acetate   Isopropyl alcohol

▲ **FIGURE 9.2** Structures of the solvents to be used for stain removal in this lab.

EXPERIMENT

Part 1. Testing Stain Removers

1. Obtain twelve 2″×2″ squares of white cotton fabric. Using a laundry or permanent marker, give each fabric square a number from 1–12. Allow the label to dry.

2. Use a blue or black ballpoint pen to draw an X (about 2 cm × 2 cm) on squares 1–3.

3. Place two drops of chocolate syrup on squares 4–6 and use your finger to spread the syrup in a circular pattern. Allow the syrup to absorb into the fabric.

4. Place two drops of cooking oil on squares 7–9 and allow the oil to absorb into the fabric.

5. Use a red lipstick to draw an X (about 2 cm × 2 cm) on squares 10–12.

6. Use Table 9.1 on the report sheet to guide your tests. As Table 9.1 demonstrates, for stained fabric squares 1–3 (those stained with ink), square 1 will be tested with hexane, square 2 with ethyl acetate, and square 3 with isopropyl alcohol. Complete the table using this same sequence with each set of stained fabric squares. As you work, take care to keep the solvents away from the labels you wrote on each fabric square as the solvents may dissolve the ink used to write the number or letter.

7. To test the stain-removing abilities of each of the solvents (hexane, ethyl acetate, and isopropyl alcohol), use the following procedure. (CAUTION: Hexane, ethyl acetate, and isopropyl alcohol are flammable and may irritate skin. Keep away from open flames and wear gloves when handling. Work in a fume hood as directed by your instructor.)

   a. Place the stained fabric to be tested on a double layer of paper towels.

   b. Using a disposable pipette, add drops of solvent to the stained area until the stain is completely wet with the solvent.

   c. Using a white paper towel, blot the wetted area to attempt to remove the stain. Do not wipe or scrub the stained area.

   d. Repeat the wetting and blotting with the same solvent.

   e. Dry the fabric square with a heat gun set on low or a stream of air.

   f. Record your observations in Table 9.1. (Note: For your observations, consider how well the stain was removed, the condition of the fabric after stain removal, etc. Ask yourself, "Would an article of clothing that had been cleaned in this manner be suitable to wear in a professional setting?")

Part 2. Comparison with Soap Solution

1. Obtain four $2'' \times 2''$ squares of white cotton fabric. Using a laundry or permanent marker, label one square with an "I" for ink, one square with a "C" for chocolate, one square with an "O" for oil, and the final square with an "L" for lipstick.

2. Prepare a solution of soap water by dissolving about a teaspoon of soap flakes in 200 mL of water.

3. Use this soap solution to wash each of the stained fabric squares. Rinse the washed square with clean water and repeat.

4. Dry each fabric square with a heat gun set on low or a stream of air.

5. For each type of stain, compare the stain-removing ability of the soap solution with the stain-removing ability of each of the solvents. Record your observations in Table 9.2.

Part 3. Combining Cleaning Forces

1. Based on your observations of the stain removal for each of the fabric squares in Part 1, gather the fabric squares that you thought were not cleaned by the tested method.

2. Using the previously prepared soap solution (from Part 2), wash each of these fabric squares, rinse with clean water, and repeat.

3. Dry each fabric square with a heat gun set on low or a stream of air.

4. Did washing these previously treated fabric squares with the soap solution aid in removing the stain? Create your own table in which to record your observations.

Pre-Lab Questions | 9

1. Review Section 3.7 (Electronegativity and Molecular Polarity) in your text and answer the following questions:

   a. Define electronegativity and explain how it affects the polarity of a molecule.

   b. What other factor(s) are important in determining whether a molecule is polar or nonpolar?

2. Draw the structures of hexane, ethyl acetate, and isopropyl alcohol. Label each molecule as polar or nonpolar and give the strongest attractive force present in the pure substance. (*Hint:* Figure 7.6 in your book may be helpful in determining attractive forces.)

3. *Like dissolves like* is the golden rule of solubility. In your own words, explain what this means as it applies to removing stains from clothing.

Name _____

Date _____ Lab Section _____

What Are the Best Stain Removers? | 9

Part 1. Testing Stain Removers

TABLE 9.1 Observations of Stain-Removing Capabilities of Various Solvents

Stain/Solvent	Hexane	Ethyl acetate	Isopropyl alcohol
Ink			
Chocolate syrup			
Cooking oil			
Lipstick			

Part 2. Comparison with Soap Solution

Record your observations of the stain-removing capabilities of soap solution on the four different stains as compared to each of the solvents listed. In your observations, record whether the soap performed better than, worse than, or similar to the indicated solvent.

TABLE 9.2 Comparing Stain-Removing Abilities of Soap Solution to Various Solvents

Stain/Solvent	Soap solution versus		
	Hexane	Ethyl acetate	Isopropyl alcohol
Ink			
Chocolate syrup			
Cooking oil			
Lipstick			

Part 3. Combining Cleaning Forces

On a separate sheet of paper, create a table in which you record your observations of the stain-removing abilities of solvent + soap solution for those stains that were not removed by solvent alone.

QUESTIONS

1. Next to each stain listed below, write the name of the solvent that was most effective in removing the stain. (Note: The stain may not have been COMPLETELY removed by any solvent; simply indicate which solvent was most effective.)

    Ink: Chocolate syrup:

    Cooking oil: Lipstick:

2. Based on your answers to question 1, would you say that the component molecules of each stain are polar or nonpolar? Explain your reasoning in the space provided.

Ink:

Chocolate syrup:

Cooking oil:

Lipstick:

3. Were any of the stains removed more effectively by combining cleaning forces—using the stain remover followed by the soap solution? If so, what can you infer about the molecular makeup of that stain?

4. Given your experience in lab today,

   a. How would you answer our guiding question: "What are the best stain removers?"

   b. Outline a strategy that you could use in the future for removing stains.

Which Triglycerides Make the Best Soap? | 10

LEARNING GOALS

After completing this laboratory exercise, you should be able to:

- Draw the structure of a triglyceride and identify its functional groups
- Explain the chemical process used to make soap
- Predict the products of a saponification reaction

BEFORE YOU BEGIN

You first encountered fatty acids as examples of nonpolar compounds in Chapter 4 of your text (*General, Organic, and Biological Chemistry, 2nd edition*). In Chapter 7, you discovered how to predict the solubility of fats and fatty acids and how to use these compounds to make soap.

Before you begin this experiment, review the material from Chapters 4, 5, and 7, focusing on the following sections and concepts:

- Section 4.3 and 4.5: Saturated and unsaturated fatty acids
- Section 5.5: Reactions of organic compounds, particularly hydrolysis reactions
- Section 7.3: Predicting solubility of a compound based on its polarity, fatty acids as an example of amphipathic compounds, micelles, and how soap works
- Section 7.5: Structures of fats and oils

INTRODUCTION

In Lab 7, you saw how condensation reactions can be used beneficially in the synthesis of pharmaceuticals. The reverse reaction of condensation is hydrolysis. Literally *breaking with water*, hydrolysis reactions break down large or complex molecules into their simpler component molecules through the addition of water. Hydrolysis reactions typically involve the addition of an acid or a base to facilitate the reaction. In today's experiment, you will utilize a base-promoted hydrolysis reaction to synthesize soap.

Soapmaking is an ancient process that typically involves combining animal fat with a strong base such as caustic potash (KOH) or lye (NaOH). Chemists call this process saponification—simply, a hydrolysis reaction that is facilitated by the addition of a base. As shown in the general reaction, Scheme 10.1, the larger and more complex triglyceride is broken down into its simpler components, including glycerol and three fatty acid salts. Fats and oils are triglycerides, which are "triple esters" of glycerol and three long-chain fatty acids.

▲ SCHEME 10.1 Saponification reaction of a triglyceride. The letter *R* represents the long hydrocarbon chains of the fatty acids. The R groups may be the same, or they may be different.

While animal fats have long been used in the soapmaking process, a variety of triglycerides, including plant-derived oils, can be used. In today's experiment, you and your classmates will choose from a variety of different fats and oils as the starting material to create your "designer" soap. You will then compare soaps to determine which fats or oils yield the hardest or softest soap, which have the most pleasant smell, and which yield the best cleaning soaps.

EXPERIMENT

Part 1. Synthesis of Soap

1. Set up a hot water bath as follows: Fill a 600 mL beaker half full of water. Place the beaker on a hot plate, add a couple of boiling chips to the water, and heat until the water is gently boiling. Make sure the beaker does not boil dry.
2. Pour approximately 100 mL of saturated NaCl (sodium chloride) solution into a 250 mL beaker and place the beaker in an ice bath to chill the NaCl solution. The cold NaCl solution will be used to help precipitate the salt from the solution.
3. Using a graduated cylinder, measure 40 mL of 6 M NaOH (sodium hydroxide) solution and pour this into a 250 mL Erlenmeyer flask. (CAUTION: Sodium hydroxide is caustic. Wear gloves and protective clothing. Handle with care. In case of spills follow clean up instructions provided by your instructor.)
4. Using the same graduated cylinder, measure 40 mL of 95% ethanol and pour this into the same 250 mL Erlenmeyer flask. Stir the resulting solution carefully with a glass stirring rod until it is mixed completely.
5. Choose one of the fat sources provided for this lab and weigh out about 10 g of the chosen fat. Place the measured sample into a 250 mL beaker.
6. Measure 40 mL of the NaOH/ethanol solution you prepared in steps 3 and 4 and pour it into the 250 mL beaker containing the fat.
7. Place the 250 mL beaker containing the fat and NaOH/ethanol solution into the boiling water bath. Make sure the beaker remains upright in the water bath and that the water from the water bath does not splash into the beaker containing the fat.
8. Stir the mixture in the beaker with a glass stirring rod.
9. Continue to heat and stir the mixture for 45 minutes. Add additional NaOH/ethanol solution to maintain a constant volume of liquid in the beaker.
10. Turn off the hot plate, remove the beaker from the hot water bath, and place it on the bench top. Allow the reaction mixture to cool to room temperature. (CAUTION: The beaker will be very hot! Wear thermal gloves or use beaker tongs to remove the beaker from the water bath.)
11. Pour all of the chilled NaCl solution that you prepared in step 2 into the beaker containing the reaction mixture and stir vigorously with a glass stirring rod. The precipitate that forms is soap.
12. Set up a vacuum filtration apparatus (as you did in Lab 7) with a Büchner funnel and appropriately sized filter paper and filter the reaction mixture.
13. Wash the solid on the filter paper with no more than 50 mL of the ice water from the ice bath used in step 2.
14. Leave the vacuum source turned on to draw air over the soap to help dry it.

Part 2. Testing Your Soap

1. After your soap has dried for 15 minutes, remove the Büchner funnel from the sidearm flask and discard all solutions as directed by your instructor.
2. Observe the properties of your soap (color, texture, and smell) and record those observations on your report sheet. (CAUTION: Your soap sample may contain unreacted base. Wear gloves to handle your soap.)
3. Find two of your classmates who used a different fat source for their soap synthesis and record the same observations of their soaps. Indicate the original fat source of each of the soaps you observe. (In your comparison, you should include at least one soap sample synthesized from an oil and one synthesized from a fat.)
4. Place a small amount of your original fat source onto a watch glass and spread it into a thin layer.
5. Test the cleaning effectiveness of your soap by using a small bit of it to wash the fat from the watch glass. Record your observations on your report sheet.
6. Repeat the cleaning test with the commercial soap provided in the lab, using the original fat source of your soap. Compare the cleaning effectiveness of the commercial soap to the cleaning effectiveness of your soap. Record your observations on your report sheet.
7. Repeat the cleaning test with the soaps of two of your classmates. For each soap you test, use the original fat source of that soap to "dirty" the watch glass.

Pre-Lab Questions | 10

1. Using Tables 4.4 and 4.7 from your textbook, draw the structure of a triglyceride containing two linoleic acids and one stearic acid.

2. Circle and name all of the functional groups in the following compound.

3. Draw the structures of the products of the following hydrolysis reaction.

$$CH_3CH_2COCH_2CH_3 \quad \xrightarrow[\text{H}_2\text{O}]{\text{NaOH}}$$

REPORT SHEET | LAB

Which Triglycerides Make the Best Soap? | 10

Part 2. Testing Your Soap

Fat source used to synthesize your soap   _____

Observations of your soap (include color, texture, and smell):

Observations of the cleaning effectiveness of your soap:

Fat source used to synthesize first comparison soap   _____

Observations of first comparison soap (include color, texture, and smell):

Observations of the cleaning effectiveness of first comparison soap:

Fat source used to synthesize second comparison soap   _____

Observations of second comparison soap (include color, texture, and smell):

Observations of the cleaning effectiveness of second comparison soap:

QUESTIONS

1. Did the fat source used to synthesize the various soaps make a difference in the texture of the soap? Were the soaps made from oils softer or harder than soaps made from fats?

2. How did the cleaning effectiveness of your soap compare to the commercial soap?

3. Of the soap samples you compared, was one of them more effective at cleaning the dirty watch glass than the others? If so, what fat source was used to synthesize that soap?

4. You have decided to open a soap company based on your experience in lab today. What steps will you take to improve the quality and marketability of the soap your company will produce?

What Is the Difference Between Osmosis and Dialysis? | 11

LEARNING GOALS

After completing this laboratory exercise, you should be able to:

- Differentiate between hypotonic and hypertonic solutions
- Explain the process of osmosis
- Predict the direction of water flow across a semipermeable membrane separating solutions of differing concentrations
- Describe the process of dialysis
- Explain the difference between osmosis and dialysis

BEFORE YOU BEGIN

You were first introduced to membranes in Chapter 7 of your text (*General, Organic, and Biological Chemistry, 2nd edition*). In Chapter 8, you saw the role that membranes play in osmosis and dialysis.

Before you begin this experiment, review the material from Chapters 7 and 8, focusing on the following sections and concepts:

- Section 7.6: The structure of the cell membrane
- Section 8.1: Solutions, solutes, and solvents
- Section 8.6: Osmosis, diffusion, and dialysis

INTRODUCTION

In Chapter 7, you learned that the membrane around living cells is a bilayer of phospholipids with polar faces on both surfaces and a nonpolar interior. The structure of the bilayer results in a semipermeable membrane that allows some substances to pass freely across the membrane but inhibits the passage of others.

When a semipermeable membrane such as the cell membrane separates two aqueous solutions that have different concentrations of solute, water molecules move across the membrane to attempt to equalize the concentrations in a process called osmosis. The water molecules move from an area of low concentration of solute (more water) to an area of high concentration of solute (less water).

If the solution inside the membrane has a higher concentration of solute than its surrounding environment, the outside solution is said to be hypotonic. In this case, there are a greater number of water molecules on the outside of the membrane, and the water will flow from the outside into the area surrounded by the membrane. If the solute concentrations are opposite those just described (higher outside the membrane than inside), the solution outside the membrane is said to be hypertonic. In this case, water will flow out of the area surrounded by the membrane into the environment. As Figure 11.1 shows, water moves across the membrane toward the side with fewer water molecules.

A simple trick can help you remember hypotonic and hypertonic. For hyp<u>O</u>tonic, there are more water molecules <u>O</u>utside the membrane (note both O's). Hypertonic is the opposite—more water molecules inside the membrane. Keep in mind that the flow of water, as seen routinely in nature, moves from an area with more water to an area with less water.

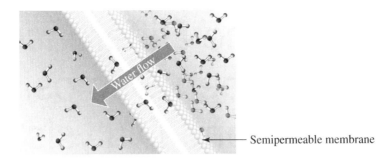

Semipermeable membrane

▲ **FIGURE 11.1** Water will flow across a semipermeable membrane from an area with more water molecules to an area with fewer water molecules.

A similar idea functions in the medically important process called dialysis, which is used to clean the blood of patients whose kidneys do not work properly. Dialysis is similar to osmosis except that dialysis involves the movement of solutes across a semipermeable membrane, whereas osmosis involves the flow of solvents, especially water, across a semipermeable membrane. During dialysis, solutes move across the membrane from an area of high solute concentration to an area of low solute concentration through a process known as diffusion. Solute molecules diffuse across the membrane to equalize the concentration on either side of the membrane. If there is no difference in concentration, no diffusion occurs.

During one type of kidney dialysis, a patient's blood is pumped into a dialysis membrane immersed in a dialyzing solution. The dialyzing solution has the same concentration of salt, glucose, and other beneficial molecules as the blood; so those molecules do not cross the membrane and therefore remain in the blood. In turn, the dialysis membrane has specifically engineered pore sizes to prevent large molecules such as proteins from crossing. Urea and other small waste molecules in the blood move easily across the dialysis membrane into the dialyzing solution, where they are subsequently discarded. The cleaned blood is then returned to the patient.

In today's experiment, you will witness the effects of osmosis on pieces of a potato placed in hypotonic and hypertonic solutions. You will also see what happens when a solution of salt, glucose, and starch is dialyzed in distilled water.

EXPERIMENT

Part 1. Observing Osmosis

1. Prepare a 5% sodium chloride (NaCl) solution as follows: Weigh 5 g of sodium chloride into a beaker or weighing boat. Measure 95 mL of water into a 100 mL graduated cylinder. Add the sodium chloride to the water in the graduated cylinder and stir to mix completely.

2. As demonstrated in Figure 11.2, obtain a baking potato and cut it in half around its circumference. From the cut end of the potato, cut a slice that is 1 cm thick. From the center of the resulting round slice, cut a rectangular section that is 1 cm wide × 4 cm long × 1 cm thick. Cut a second rectangular section of potato of the same size from the round slice.

3. Weigh each piece of potato and record the mass on the report sheet.

4. Place one of the potato pieces in a 250 mL beaker labeled "Water" and cover it completely with distilled water.

5. Place the second potato piece in a different 250 mL beaker labeled "Salt Water" and cover it completely with the 5% salt solution that you prepared previously in step 1. Save the remainder of the 5% salt solution for Part 2 of the experiment.

6. Allow the beakers to sit undisturbed for at least 30 minutes (an hour is better if time permits). Move on to Part 2 of the experiment while you are waiting.

7. Remove the potato pieces from each beaker, making sure to keep track of which was in the distilled water and which was in the salt water. Carefully and quickly blot each potato piece with a paper towel to dry it.

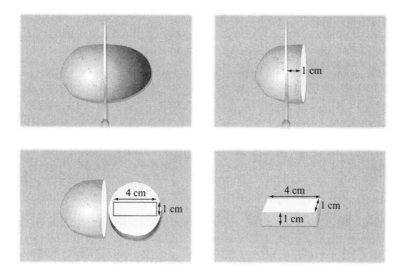

▲ **FIGURE 11.2** Cutting potato pieces for osmosis experiment.

8. Weigh each potato piece and record the mass on the report sheet.

9. Observe the appearance and texture of each potato piece and record your observations on the report sheet.

Part 2. Detecting Dialysis

1. Obtain a 15–20 cm length of dialysis tubing and soak it in a beaker of distilled water for 5–10 minutes.

2. Remove the tubing from the beaker and tie a knot in one end so that liquid cannot escape. Open the other end of the tube, taking care not to tear it.

3. Insert a funnel into the opening of the tube and add 1% starch solution until the tube is one-third to one-half full. Add 10% glucose solution until the tube is about three-quarters full. Add 5% sodium chloride solution (left over from Part 1) until the tube is almost full. Leave enough room at the top of the tube to tie it.

4. Twist the top of the tube several times to close it and tie it tightly with a piece of string or secure with a clamp.

5. Rinse the outside of the filled tube thoroughly with distilled water to clean it of any spills.

6. Place the dialysis tubing into a 500 mL beaker and cover it completely with distilled water.

7. Allow the tubing to sit undisturbed in the distilled water for 30 minutes.

8. While you are waiting, read the Procedures for the Chloride, Glucose, and Starch Detection Tests, which follows, and prepare your materials.

9. After 30 minutes, remove the dialysis tube from the water and set it aside in a clean, dry beaker.

10. Measure 15 mL of the dialysate (the water in the beaker that was used for dialysis) into a 25 or 50 mL graduated cylinder.

11. Label three test tubes as follows: CW (Chloride, Water), GW (Glucose, Water), and SW (Starch, Water).

12. Measure 5 mL of the dialysate from the graduated cylinder into each of the test tubes.

13. Using the following Procedures for Chloride, Glucose, and Starch Detection Tests, test the CW tube for the presence of chloride, the GW tube for the presence of glucose, and the SW tube for the presence of starch. Record your observations on the report sheet.

14. Untie the dialysis tubing and pour 15 mL of the solution from inside the tubing into a clean 25 or 50 mL graduated cylinder.

15. Label three test tubes as follows: CT (Chloride, Tubing), GT (Glucose, Tubing), and ST (Starch, Tubing).

16. Measure 5 mL of the solution from the graduated cylinder into each of the test tubes.

17. Using the following Procedures for Chloride, Glucose, and Starch Detection Tests, test the CT tube for the presence of chloride, the GT tube for the presence of glucose, and the ST tube for the presence of starch. Record your observations on the report sheet.

18. Dispose of all solutions and reagent strips as directed by your instructor.

PROCEDURES FOR CHLORIDE, GLUCOSE, AND STARCH DETECTION TESTS

Each of the following tests should be conducted alongside a blank (5 mL of distilled water in place of solution to be tested) to show the results of a negative test. If your instructor directs, you may conduct each test with the original solutions (5% sodium chloride, 10% glucose, and 1% starch) to witness the results of a positive test.

a. Chloride test: Place 5 mL of solution to be tested in a test tube, add 3–4 drops of $AgNO_3$ Reagent, stopper the test tube, hold the top of the tube between your thumb and forefinger, and tap the bottom of the tube to mix. (CAUTION: $AgNO_3$ will stain your skin. Wear gloves and handle with care.) The formation of a white precipitate (solid) is a positive test and indicates the presence of chloride ion.

b. Glucose test: Place 5 mL of the solution to be tested in a test tube, dip the test end of a Keto-Diastix Regeant Strip into the solution, and remove immediately. Wait 30 seconds; then compare the glucose test area on the strip to the Glucose Color Chart (on the bottle). A change in color of the glucose test area on the test strip is a positive test and indicates the presence (and concentration) of glucose. Note: This test is time-sensitive and must be read at 30 seconds.

c. Starch test: Place 5 mL of the solution to be tested in a test tube, add 3–4 drops of I_2/KI Reagent, stopper the test tube, hold the top of the tube between your thumb and forefinger, and tap the bottom of the tube to mix. The formation of a deep blue-black color is a positive test and indicates the presence of starch.

Name _____

Date _____ Lab Section _____

Pre-Lab Questions | 11

1. Assuming that the concentration of oxygen in the water is the same, why does a freshwater fish die if it is released into the ocean?

2. If a raisin is placed in a glass of tap water, what will happen to the raisin? Explain what happens in this process.

3. Explain why substances such as sodium ions and chloride ions do not move freely across our cell membranes.

4. In terms of concentration, what direction do solute molecules move during diffusion?

REPORT SHEET | LAB

What Is the Difference Between Osmosis and Dialysis? | 11

Part 1. Observing Osmosis

Mass of potato section placed in distilled water   _____ g

Mass of potato section after removing from distilled water   _____ g

Difference in mass   _____ g

Mass of potato section placed in salt water   _____ g

Mass of potato section after removing from salt water   _____ g

Difference in mass   _____ g

Observations

Appearance and texture of potato after removing from distilled water

Appearance and texture of potato after removing from salt water

Part 2. Detecting Dialysis

For each of the tests listed, indicate whether the test was positive or negative for the indicated substance.

Chloride test of CW   _____

Glucose test of GW   _____

Starch test of SW   _____

Chloride test of CT   _____

Glucose test of GT   _____

Starch test of ST   _____

QUESTIONS

1. Consider the potato section that was left in distilled water:

   a. Did it gain or lose mass? Explain the change.

   b. Was the water hypotonic or hypertonic to the potato? Explain your answer.

   c. Compare the texture of the potato before it was placed in distilled water with its texture afterwards. Was the potato more firm or less firm after being in the water? How do you explain this change?

2. Consider the potato section that was left in salt water:

   a. Did it gain or lose mass? Explain the change.

   b. Was the salt water hypotonic or hypertonic to the potato? Explain your answer.

   c. Compare the texture of the potato before it was placed in salt water with its texture afterwards. Was the potato more firm or less firm after being in the salt water? How do you explain this change?

3. What substances diffused through the dialysis tubing? What substances did not diffuse through the dialysis tubing? How do you explain the difference?

4. Of the substances that did diffuse through the dialysis tubing, did all of the molecules/ions of that substance diffuse into the dialysate? Explain.

5. Explain the difference between osmosis and dialysis.

Which Antacid Is Most Effective at Relieving Heartburn? | 12

LEARNING GOALS

After completing this laboratory exercise, you should be able to:

- Discuss the pH scale and give examples of acidic and basic compounds
- Explain how antacids affect the pH of stomach acid
- Demonstrate how to measure pH of a solution and explain the significance of the value obtained

BEFORE YOU BEGIN

In Chapter 9 of your text (*General, Organic, and Biological Chemistry, 2nd edition*), you were introduced to acids and bases and discovered how they react with each other. You also discovered how to use the pH scale to determine the acidity of a substance.

Before you begin this experiment, review the material from Chapter 9, focusing on the following sections and concepts:

- Section 9.1: Definition of acid and base
- Section 9.2: Acid-base neutralization reactions and antacids
- Section 9.5: Definition of pH and the pH scale and calculation of pH and acid concentration

INTRODUCTION

Acidic and basic compounds have a wide variety of useful applications. Bases are cleansers such as ammonia and soaps, and drain openers. Acids are ingredients in foods and components of food preparation processes such as pickling. They are particularly useful in digestive processes where acids play a key role, especially in the digestion of proteins. The initial digestion of protein occurs in the gastric juices of the stomach, which are composed mainly of hydrochloric acid and have a pH between 1.5–3.5. The gastric juices are secreted by the parietal cells that line the stomach walls. Certain stimuli such as stress and particular types of food cause the parietal cells to overproduce acid. The result is excess acid in the stomach, which may splash into the esophagus, producing a condition commonly known as heartburn or acid indigestion.

Over-the-counter (OTC) medicines called antacids are the most common remedies for occasional heartburn. (Medicines for chronic heartburn function much differently and are not considered in today's lab.) Antacids work by neutralizing excess acid produced by the stomach to help relieve the pain and burning sensation often associated with heartburn. The majority of antacids are weak or slightly soluble bases. Compounds that are strongly basic are not used as antacids because they would damage the mucous lining of the stomach and create problems much worse than heartburn. Table 12.1 lists some common antacids and their active ingredients.

TABLE 12.1 Common Antacids

Antacid	Active Ingredient(s)
Milk of Magnesia	Magnesium hydroxide
Tums®	Calcium carbonate
Rolaids®	Calcium carbonate and magnesium hydroxide
Baking soda	Sodium bicarbonate
Maalox®	Aluminum hydroxide and magnesium hydroxide

In today's lab, you will test the effectiveness of three different antacids—magnesium hydroxide, the active ingredient in milk of magnesia; calcium carbonate, the active ingredient in Tums®; and sodium bicarbonate, or baking soda, which is a common home remedy for heartburn. The balanced equations for the reactions of the antacids with stomach acid (hydrochloric acid) are given below.

$$Mg(OH)_2 + 2HCl \rightarrow MgCl_2 + 2H_2O$$

$$CaCO_3 + 2HCl \rightarrow CaCl_2 + CO_2 + H_2O$$

$$NaHCO_3 + HCl \rightarrow NaCl + CO_2 + H_2O$$

You will use hydrochloric acid, the acid in gastric juices, at a concentration of 0.10 M (pH = 1) to simulate the environment of a stomach with excess acid. For the first part of the experiment, you will use pure forms of the active ingredients of the antacids, and in Part 2, you will test the effectiveness of the actual antacids you would purchase at your local pharmacy.

EXPERIMENT

For the first part of the experiment, you will use the pure compounds, which serve as the active ingredients in the antacids you will study in Part 2.

Part 1. Effect of Antacid on pH of Stomach Acid

1. Obtain three 250 mL beakers and using a marker, label one magnesium hydroxide, one calcium carbonate, and one sodium bicarbonate.

2. Using a graduated cylinder, measure and pour 100 mL of 0.10 M hydrochloric acid into each beaker. (CAUTION: Hydrochloric acid is corrosive. Wear gloves, safety glasses, and protective clothing. Handle with care.)

3. Using a pH wand or pH meter (as demonstrated by your instructor), measure the pH of the acid solution in one of the beakers and record that value on the report sheet as pH of "stomach acid".

4. Using a laboratory balance, measure as nearly as possible 0.25 g each of magnesium hydroxide, calcium carbonate, and sodium carbonate onto separate watch glasses or weighing papers. Record the actual mass on the report sheet.

5. Working with one beaker at a time, carefully and slowly add the solid into the beaker labeled for it. Stir the solution carefully with a stirring rod. (CAUTION: Solutions may become hot or may fizz. Wear gloves, safety glasses, and protective clothing.) Record your observations of each reaction on the report sheet.

6. Stir each solution for 3–5 minutes. If the solution becomes hot, allow it to cool to room temperature before proceeding. You may run cool water over the outside of the beaker to speed the cooling process.

7. Using a pH wand or pH meter, measure the pH of each solution and record that value on the report sheet.

8. Dispose of each solution as directed by your instructor.

Part 2. Measuring Effectiveness of Antacids per Dose

For this part of the experiment, you will use the commercial antacids for your tests.

1. Obtain three 250 mL beakers and label each beaker with the brand name of one of the antacids. Each beaker should be labeled with a different antacid.

2. Using a graduated cylinder, measure and pour 100 mL of 0.10 M hydrochloric acid into each beaker. (CAUTION: Hydrochloric acid is corrosive. Wear gloves, safety glasses, and protective clothing. Handle with care.)

3. Using a pH wand or pH meter (as demonstrated by your instructor), measure the pH of the acid solution in one of the beakers and record that value on the report sheet as pH of "stomach acid".

4. Check with your instructor to determine the amount of antacid equal to one adult dose and the identity of the active ingredient. Obtain that amount of each of the three antacids.

5. If the antacid is a tablet or pill, use a mortar and pestle to grind it to a fine powder. If the antacid is a liquid, you will use the liquid form.

6. Working with one beaker at a time, carefully and slowly add the single dose of the antacid into the beaker labeled for it. Stir the solution carefully with a stirring rod. (CAUTION: Solutions may become hot or may fizz. Wear gloves, safety glasses, and protective clothing.) Record your observations of each reaction on the report sheet.

7. Stir each solution for 3–5 minutes. If the solution becomes hot, allow it to cool to room temperature. You may run cool water over the outside of the beaker to speed the cooling process.

8. Using a pH wand or pH meter, measure the pH of each solution and record that value on the report sheet.

9. Dispose of each solution as directed by your instructor.

Pre-Lab Questions | 12

1. Refer to Chapter 3 of your text as needed and write the chemical formulas for each of the following compounds:

   a. magnesium hydroxide   _____

   b. calcium carbonate   _____

   c. sodium bicarbonate   _____

2. Hydrochloric acid (HCl) is a strong acid, and sodium hydroxide (NaOH) is a strong base.

   a. Write an equation showing the reaction between these two compounds.

   b. What type of reaction takes place between HCl and NaOH?

3. For each of the following compounds, decide if it is acidic, basic, or neutral.

   a. soap, pH = 10   _____

   b. urine, pH = 6   _____

   c. ammonia, pH = 11.5   _____

   d. pure water, pH = 7   _____

   e. lemon juice, pH = 2   _____

4. Explain why strong bases are not used as antacids.

REPORT SHEET │ LAB

Which Antacid Is Most Effective at Relieving Heartburn?

│ **12**

Part 1. Effect of Antacid on pH of Stomach Acid

Initial pH of "stomach acid" in beakers   _____

Mass of magnesium hydroxide used   _____ g

Mass of calcium carbonate used   _____ g

Mass of sodium bicarbonate used   _____ g

Final pH of stomach acid + magnesium hydroxide   _____

Final pH of stomach acid + calcium carbonate   _____

Final pH of stomach acid + sodium bicarbonate   _____

Observations

Describe the reaction that occurred in each beaker after the addition of the antacid compound.

Part 2. Measuring Effectiveness of Antacids per Dose

Initial pH of "stomach acid" in beakers   _____

Brand name of antacid 1   _____

Active ingredient of antacid 1   _____

Amount of active ingredient in one adult dose of antacid 1   _____ g

Brand name of antacid 2   _____

Active ingredient of antacid 2   _____

Amount of active ingredient in one adult dose of antacid 2   _____ g

Brand name of antacid 3   _____

Active ingredient of antacid 3   _____

Amount of active ingredient in one adult dose of antacid 3   _____ g

Final pH of stomach acid + antacid 1   _____

Final pH of stomach acid + antacid 2   _____

Final pH of stomach acid + antacid 3   _____

Observations

Describe the reaction of each antacid with the stomach acid in the beaker.

QUESTIONS

1. Considering the reactions of Part 1,

   a. Which of the compounds changed the pH of the stomach acid the most?

   b. Which of the compounds changed the pH of the stomach acid the least?

   c. Based on this initial test, which compound was most effective (per 0.25 g) at neutralizing the stomach acid?

2. Considering the reactions of Part 2,

   a. Did any of the antacids change the pH of the stomach acid more than the others? If so, which one produced the greatest change?

   b. Based on this test, which (if any) of the antacids was most effective (per dose) at neutralizing the stomach acid?

3. Considering your results from Parts 1 and 2, which antacid would you recommend as being most effective? Explain your reasoning.

Where Is Lactose Digested? | 13

LEARNING GOALS

After completing this laboratory exercise, you should be able to:

- Identify the substrate of lactase and the products of the reaction it catalyzes
- List at least two factors that affect an enzyme's activity
- Describe the process used to determine an enzyme's optimum pH

BEFORE YOU BEGIN

In Chapter 10 of your text (*General, Organic, and Biological Chemistry, 2nd edition*), you were introduced to enzymes and the factors that affect the reactivity of an enzyme. Before you begin this experiment, review the material indicated here focusing on the following sections and concepts:

- Section 6.5: Disaccharides and structures of lactose, sucrose, and maltose
- Section 10.6: Enzymes as functional proteins, enzyme activity, including active site and enzyme–substrate models
- Section 10.7: Factors that affect enzyme activity, including pH and temperature

INTRODUCTION

Lactase is the enzyme responsible for digestion of milk sugar (lactose) in our bodies. It functions by hydrolyzing the disaccharide lactose into its monosaccharide components—galactose and glucose. Lactose intolerance is a medical condition characterized by the onset of abdominal cramps and diarrhea following ingestion of foodstuffs containing lactose and is caused by the lack of lactase in the digestive tract. This condition is virtually nonexistent in infants, but for reasons not entirely clear to medical science, it becomes more prevalent after childhood. People who suffer from this condition must refrain from the consumption of dairy products or take dietary supplements of the missing enzyme so that they can digest the lactose present in these foods.

In this experiment, you will use a commercially available source of lactase to study the hydrolytic action of the enzyme. Given that glucose is one of the products of the reaction catalyzed by lactase, you will use commercially available urinalysis test strips to detect the presence of glucose in the test solutions. These test trips are used in home health situations to accurately determine the concentration of glucose present in a urine sample of a diabetic patient.

We know that lactose, a glucose-containing disaccharide, is a substrate for lactase. To determine the specificity of the enzyme, you will combine lactase with two other glucose-containing disaccharides separately and test each of those solutions for enzyme activity by using the urinalysis strips to detect the presence of the monosaccharide glucose.

Finally, you will determine the optimum pH for lactase by studying its action at three different pHs; then you will use this information to determine where lactase is located along the digestive tract. The digestive enzymes of the human body are distributed throughout the digestive tract. A few, such as amylase, are located in the mouth, a region of nearly neutral pH (6–7); some are located in the stomach, a region of acidic pH (1); some are located in the first few inches of the small intestine (the duodenum), also a region of neutral pH; and some are located further along in the intestines, a region of slightly basic pH (8). By varying the pH of your test solutions to include pHs of 1, 7, and 8 and monitoring the activity of lactase in each, you should be able to determine whether lactase is located in the stomach, in the duodenum, or further along in the intestines.

EXPERIMENT

Part 1. Preparation of the Lactase Solution

1. Obtain 1 Lactaid pill and grind it up using a mortar and pestle.
2. In a small beaker, combine the crushed pill and 10 mL of distilled water. Swirl the solution for approximately 5 minutes.
3. Decant the liquid into a clean beaker and discard the remaining solid as directed by the instructor.
4. Use a pH meter or a pH wand as directed by the instructor to determine the pH of this solution and record that pH on your report sheet.
5. Label this solution "Lactase Enzyme Prep" and set in a safe place for use throughout the experiment.

Part 2. Studying the Specificity of Lactase

1. Obtain 10 mL of a 1% lactose solution and place it in a small Erlenmeyer flask or beaker.
2. Add 1 mL of your lactase enzyme prep to the flask or beaker.
3. Immediately place the flask or beaker in a water bath at 37°C and start your timer (or note the time on your watch and record it on your report sheet).
4. Carefully swirl the reaction solution in the flask or beaker every 5 minutes. Try to minimize time out of the water bath while swirling.
5. When 15 minutes have elapsed, use the Keto-Diastix Regeant Strips to determine the concentration of glucose present in the solution. Record this value on your report sheet.
6. Use the Keto-Diastix Regeant Strips to determine the concentration of glucose present in the solution at 30 minutes and 45 minutes. Record these values on your report sheet.
7. Discard the solutions down the drain with plenty of water and place the used test strips in the trash for both solution and strip or as directed by the instructor.
8. Repeat steps 2–7 using 10 mL of a 1% sucrose solution. Record all data on your report sheet.
9. Repeat steps 2–7 using 10 mL of a 1% maltose solution. Record all data on your report sheet.
10. Prepare a graph with "Glucose Concentration" along the y-axis (vertical) and "Time" along the x-axis (horizontal). Use a different color (or different type data point) for each of the disaccharides. For each individual data set, draw the best straight line through the data points.

Part 3. Studying the Optimum pH of Lactase

1. Obtain 10 mL of a 1% lactose solution and place it in a small Erlenmeyer flask or beaker.
2. Add 2 drops of 1.0 M HCl (hydrochloric acid) to the flask or beaker and quickly determine the pH of the solution using a pH meter or pH wand. The pH should be approximately 1. If it isn't, add 1.0 M HCl dropwise until pH is approximately 1. (CAUTION: Hydrochloric acid is corrosive. Wear gloves, safety glasses, and protective clothing. Handle with care.)
3. Add 1 mL of your lactase enzyme prep to the flask or beaker.
4. Immediately place the flask or beaker in a water bath at 37°C and start your timer (or note the time on your watch and record it on your report sheet).
5. Carefully swirl the reaction solution in the flask or beaker every 5 minutes. Try to minimize time out of the water bath while swirling.
6. When 15 minutes have elapsed, use the Keto-Diastix Regeant Strips to determine the concentration of glucose present in the solution. Record this value on your report sheet.
7. Use the Keto-Diastix Regeant Strips to determine the concentration of glucose present in the solution at 30 minutes and 45 minutes. Record these values on your report sheet.
8. Discard the reaction solutions as directed by your instructor and place the used test strips in the trash or as directed by the instructor.
9. Repeat steps 1–8, but use a 0.1 M NaOH (sodium hydroxide) solution (in step 2) to adjust the pH to approximately 8. Record all data on your report sheet. (CAUTION: Sodium hydroxide is caustic. Wear gloves, safety glasses, and protective clothing. Handle with care.)
10. Prepare a graph with "Glucose Concentration" along the y-axis (vertical) and "Time" along the x-axis (horizontal). Use a different color (or different type data point) for data collected at each pH value. Use the data from Part 2 of the experiment (data collected when lactose was used as substrate) for a neutral pH. For each individual data set, draw the best straight line through the data points.

Pre-Lab Questions | 13

1. Explain the difference between the lock-and-key and induced-fit models of enzyme activity.

2. What is the medical condition known as lactose intolerance? Explain what causes it and describe its symptoms.

3. Draw the structures of the disaccharides lactose, sucrose, and maltose. Circle the glucose portion of each.

4. Enzymes have an optimum pH at which they function best.

   a. What happens to an enzyme when it is placed in an environment that is well above or well below its optimum pH? (*Hint:* See Section 10.4 of the text.)

   b. Why do these changes affect the enzyme's activity?

Name _____

Date _____ Lab Section _____

REPORT SHEET | LAB

Where Is Lactose Digested? | 13

Part 1. Preparation of the Lactase Solution

pH of lactase enzyme prep solution   _____

Part 2. Studying the Specificity of Lactase

Lactase + Lactose

    Reaction start time   _____

Lactase + Sucrose

    Reaction start time   _____

Lactase + Maltose

    Reaction start time   _____

TABLE 13.1 Concentrations of Glucose from Reaction of Lactase and Three Disaccharides

Reaction with:	Glucose Concentration at:		
	15 min	**30 min**	**45 min**
Lactose			
Sucrose			
Maltose			

Use the data in Table 13.1 to prepare the graph as directed. Attach your graph to this report sheet.

Part 3. Studying the Optimum pH of Lactase

Reaction at pH 1

    Reaction start time   _____

Reaction at pH 8

    Reaction start time   _____

Reaction at pH 6–7 (neutral, take data from Part 2, Lactase + Lactose above)

    Reaction start time   _____

TABLE 13.2 Concentrations of Glucose from Reaction of Lactase and Lactose at Three pHs

Reaction at:	Glucose Concentration at:		
	15 min	30 min	45 min
pH 1			
pH 8			
pH 6–7*			

*Copy data from Table 13.1 in Part 2, Lactase + Lactose

Use data in Table 13.2 to prepare the graph as directed. Attach your graph to this report sheet.

QUESTIONS

1. Based on the data and graph from Part 2, would you say that lactase has one and only one substrate or is it able to accommodate different substrates? Draw on the information from your graph to explain your answer.

2. The line with the greatest slope on the graph that you constructed using the data from Part 3 represents the pH at which lactase was most active. Based on the experiments you conducted in Part 3 and your graph, what pH or pH region would you say is the optimum pH for lactase?

3. Given your answer to the previous question and the information from the introduction, where along the digestive tract does lactase function?

4. Convert 37°C to Fahrenheit? Why were all experiments in Parts 2 and 3 conducted at this temperature?

How Big Is a DNA Molecule? | 14

LEARNING GOALS

After completing this laboratory exercise, you should be able to:

- Outline the steps necessary to isolate DNA from a living cell
- Explain the size of the DNA molecule compared to other molecules

BEFORE YOU BEGIN

In Chapter 11 of your text (*General, Organic, and Biological Chemistry, 2nd edition*), you learned the structure of DNA and discovered how that structure is built from sugars, nitrogenous bases, and phosphates. You were also introduced to chromosomes and their function of storing the DNA in a cell.

Before you begin this experiment, review the material from Chapter 11, focusing on the following sections and concepts:

- Section 11.1: Components of DNA and the difference between a nucleoside and a nucleotide
- Section 11.2: Formation of a nucleic acid and the primary structure of DNA
- Section 11.3: Secondary and tertiary structure of DNA, complementary base pairs, the double helix, and chromosome structure

INTRODUCTION

The smallest thing you can see with the naked eye, unaided by any special tools or lenses, is about 0.1 mm, roughly the diameter of a human hair. A water molecule, on the other hand, has a diameter of 0.00000028 mm. As you know and as this data clearly demonstrates, you cannot see molecules with the naked eye. Or can you? In today's experiment, we will challenge this supposition by attempting to isolate one of the largest molecules in a living cell, deoxyribonucleic acid (DNA).

The genetic information necessary for life is stored in DNA, which is stored in the nucleus of eukaryotic cells. As you will see in today's experiment with onion cells, DNA is huge in comparison to other molecules; yet, it remains dissolved in the solution of the cell. Your tasks today will be to disrupt the cellular structures encapsulating the solutions containing the DNA and to isolate the DNA from the other components.

To begin the process of extracting the DNA, you will grind the onion with sand to disrupt the cell wall and membrane and release the cellular contents. You will mix the resulting "onion paste" with a detergent that will disrupt the remaining membranes and release the DNA. You will also add salt to this solution and warm it gently. The sodium ions of the salt will interact with the negatively charged phosphates along the backbone of the DNA and cause the DNA to unwind from the histones, which are small proteins that act as spools for the DNA to wind around. The heat serves to denature nuclease enzymes that would damage the DNA and to denature other proteins in the cellular solution, causing some of them to precipitate from the solution. After the mixture is filtered, the addition of cold ethanol causes the nucleic acids to precipitate and the long strands of DNA can be wound onto a glass rod.

EXPERIMENT

Solution and Sample Preparation

1. Set up a warm water bath at 60°C as follows: Fill a 400 mL beaker half full with water. Place the beaker on a hot plate and gently heat until the temperature reaches 60°C. Monitor the temperature with a thermometer and do not allow it to go above 60°C.

2. Measure 100 mL of distilled water into a 250 mL beaker and place the beaker in the warm water bath. Stir the water in the 250 mL beaker with a glass stirring rod and monitor its temperature until it reaches 60°C.

3. Using a knife, cut off a sample of onion with a mass of approximately 30 g and chop the sample into small pieces. The smaller the pieces the better, but do not chop the onion into mush.

Extraction of DNA from Onion Cells

4. Obtain a mortar and pestle from the laboratory refrigerator or cooler. Work quickly to keep the temperature of the mortar and pestle as cool as possible.

5. Add the chopped onion pieces and 20 g of sand to the mortar. Vigorously grind the onion and the sand for 5 minutes until a mushy paste is formed.

6. Add 10 g of NaCl and 10 mL of liquid dish detergent to the 60°C water in the 250 mL beaker and stir with a glass stirring rod to dissolve.

7. Measure 30 mL of the NaCl/detergent solution into a 100 mL beaker and add the onion/sand mush from the mortar to the solution. Stir with a glass stirring rod to mix. Place the beaker in the water bath and maintain its temperature at 60°C.

8. Continue stirring the mixture for 10 minutes while ensuring that the temperature remains at 60°C.

9. Place a double layer of cheesecloth into a funnel and secure the funnel to a ring stand with an iron ring or three-finger clamp. Place a 150 mL beaker under the funnel.

10. Allow the solids in the beaker to settle to the bottom. Carefully pour the liquid from the onion/sand/NaCl/detergent mixture onto the cheesecloth in the funnel. Try to leave the sand and cellular debris in the beaker. Pause as necessary and squeeze the liquid from the cheesecloth into the funnel.

11. Estimate the volume of the liquid (filtrate) and allow it to cool to room temperature.

Precipitation and Isolation of DNA

12. Obtain 1.5–2 times the volume of the filtrate of ice-cold absolute ethanol.

13. Holding the beaker containing the filtrate at an angle, slowly pour the ice-cold ethanol down the side so that it forms a layer on top of the filtrate. Add the ethanol carefully so that the filtrate is not disturbed or mixed with the ethanol.

14. Carefully return the beaker to the benchtop and allow it to sit undisturbed for 2 minutes. You should see whitish, wispy strands of DNA at the interface.

15. Insert a clean glass stirring rod just below the interface of the two solutions and rotate the stirring rod to spool the DNA onto the stirring rod as you would spaghetti onto a fork.

16. When no more DNA spools onto the stirring rod, remove it from the solution. The DNA will look like a gelatinous mass. Place it on a paper towel and allow it to dry.

17. Discard all solutions as directed by your instructor.

18. Record your observations of the DNA on your report sheet.

Pre-Lab Questions | 14

1. Where is DNA found in the cells of higher organisms (eukaryotes)?

2. What is a histone? What purpose does it serve in the cell?

3. Describe each of the levels (primary, secondary, and tertiary) of the structural hierarchy of DNA.

4. Which component of DNA gives it an overall negative charge?

REPORT SHEET | LAB

How Big Is a DNA Molecule? | 14

EXTRACTION AND ISOLATION OF DNA FROM ONION CELLS

Observations

Describe the appearance of the DNA as it forms at the solution interface and as it was spooled onto the glass rod.

Describe the appearance of the DNA after drying.

QUESTIONS

1. The sodium ions of NaCl interact with DNA and cause it to unwind from the histones. Given this, what can you infer about the overall charge of the histones?

2. Each base pair along the backbone of DNA occupies about 3.3×10^{-10} m (0.33 nm). A small strand of human DNA has approximately 220 million base pairs. How long is this DNA molecule in centimeters? How long is it in inches?

3. The diameter of the DNA double helix is approximately 2 nm. Given this diameter and the length you calculated in question 2, is it possible to see a single DNA molecule with the naked eye? Why or why not?

How Is Food Broken Down During Digestion? | 15

LEARNING GOALS

After completing this laboratory exercise, you should be able to:

- List the products of digestion of starch, protein, and fat
- Describe the chemical processes involved in digestion

BEFORE YOU BEGIN

In Chapter 12 of your text (*General, Organic, and Biological Chemistry, 2nd edition*), you saw an overview of metabolism and discovered how our foodstuffs are converted into energy. Early in the chapter, you were introduced to the enyme-catalyzed digestion of the biomolecules in foods.

Before you begin this experiment, review the material from Chapter 12, focusing on the following sections and concepts:

- Section 12.1: Metabolism and the products of digestion
- Section 12.3: Digestion of carbohydrates, fats, and proteins

INTRODUCTION

Our foods are made up of carbohydrates, proteins, and fats. These complex biomolecules are the sources of the energy needed to sustain life as well as the building blocks needed to maintain and repair our bodies. Because the biomolecules contained in our foodstuffs are highly complex molecules, they must be broken down before they can be effectively utilized in our bodies. The initial step of this breaking-down process is digestion.

The digestion of the biomolecules in our foodstuffs is catalyzed by several enzymes, each specific to a particular class of biomolecules. Regardless of the biomolecule or enzyme involved, the reactions of the digestive process are enzyme-catalyzed hydrolysis reactions. As we saw in Experiment 10, hydrolysis reactions involve the breakdown of large molecules into smaller ones through the addition of water.

In today's experiment, you will investigate the digestion of the three primary classes of biomolecules in foods—proteins, carbohydrates, and fats. For each experiment, you will conduct one or more control experiments simultaneously. The controls allow you to observe what happens to the biomolecules in the absence and presence of the enzyme. By comparing the results of the control with those of the enzyme-containing experiment, you can verify the changes that are due to the action of the enzyme.

You will begin by investigating the action of pepsin on the proteins in the white of a hard-boiled egg. Pepsin is a peptidase enzyme that hydrolyzes the peptide bond between certain amino acids (see Scheme 15.1). The control experiment in this portion of the lab involves investigation of the effect of hydrochloric acid and water on the egg white. In addition, you will observe the action of pepsin under neutral and acidic conditions.

Your investigation of carbohydrate digestion will focus on starch and the action of amylase on starch. Amylase catalyzes the hydrolysis of the α-1,4 linkages in starch (see Scheme 15.1). For the control experiment, you will simply omit the amylase from the reaction mixture and observe any changes.

For your final experiment, you will consider the digestion of the fat in heavy cream by the enzyme pancreatin. Pancreatin hydrolyzes the ester linkages of triglycerides to produce glycerol and fatty acids (see Scheme 15.1). Because fats are not water-soluble molecules, their digestion is a more complex process. The enzymes that catalyze the chemistry of life are water-soluble compounds. For an enzyme to interact with a fat molecule, the two must be in solution together. Our bodies help to solubilize fats by producing bile. Bile is an amphipathic compound, meaning it has both hydrophilic and hydrophobic parts (similar to the soap molecules you studied in Experiment 9). Bile emulsifies fat molecules by surrounding them to form a micelle, analogous to soap's action on a greasy stain. Once the fat is emulsified, the enzyme works to break it down. The controls for this experiment will allow you to investigate the action of bile by itself and pancreatin by itself.

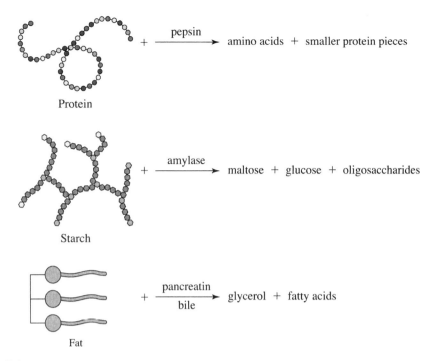

▲ **SCHEME 15.1** Digestion reactions of protein, starch, and fat.

EXPERIMENT

Part 1. Digestion of Proteins

1. Obtain a hard-boiled egg and remove and discard the shell and the yolk.

2. Use a knife to cut out three thin cubes of the remaining white of the egg. Each cube should be approximately 1 cm × 1 cm × 0.5 cm.

3. Obtain three test tubes and use a permanent marker to label them as follows: "H_2O + HCl" on the first tube, "pepsin + HCl" on the second tube, and "pepsin + H_2O" on the third tube.

4. Add 4 mL of distilled water and 1 mL of 0.1 M HCl to the first tube.

5. Add 4 mL of 2% pepsin solution and 1 mL of 0.1 M HCl to the second tube.

6. Add 4 mL of 2% pepsin solution and 1 mL of distilled water to the third tube.

7. Add 1 cube of egg white to each of the tubes.

8. Place the tubes in a 37°C water bath and allow them to sit undisturbed. Make sure the tubes are supported and will not fall over in the water bath.

9. Record your initial observations of each test tube on the report sheet.

10. After 15 minutes, note any changes and record your observations on the report sheet.

11. Repeat the observations at 15-minute intervals for an hour total time. Each time, note any changes and record your observations on the report sheet.

Part 2. Digestion of Starch

1. Obtain two test tubes and use a permanent marker to label the first tube "Amylase" and the second tube "H_2O."

2. Add 4 mL of 1% starch solution to each test tube.

3. Add 1 mL of amylase solution to the first test tube and mix thoroughly.

4. Add 1 mL of distilled water to the second test tube and mix thoroughly.

5. Record your initial observations of each test tube on the report sheet.

6. Place the tubes in a 37°C water bath and allow them to sit undisturbed. Make sure the tubes are supported and will not fall over in the water bath.

7. After 5 minutes, using a clean dropper for each test tube, remove 10 drops of the solution from each test tube and place each sample in a separate well on a spot plate.

8. Dip the test end of a Keto-Diastix® Reagent Strip into the solution from the amylase test tube and remove immediately. Wait 30 seconds; then compare the glucose test area on the strip to the Glucose Color Chart (on the bottle).

9. Repeat step 8 for the solution from the H_2O test tube.

10. Record your observations on the report sheet.

11. Add 1 drop of I_2/KI reagent to each of the samples in the spot plate and record your observations on the report sheet. (You may recall from Experiment 11 that the formation of a deep blue-black color indicates the presence of starch.)

12. Repeat steps 7–10 at 10 minutes and again at 15 minutes.

Part 3. Digestion of Fats

1. Obtain three test tubes and use a permanent marker to label them as follows: "Bile Salts" on the first tube, "Pancreatin" on the second tube, and "Pancreatin + Bile" on the third tube.

2. Add 3 mL of heavy cream (or whole milk) to each test tube.

3. Add 5 mL of distilled water and 2 mL of 5% bile salt solution to the first test tube. Cork the tube with a cork stopper and shake vigorously to mix the contents.

4. Add 5 mL of 2% pancreatin solution and 2 mL of distilled water to the second test tube. Cork the tube with a cork stopper and shake vigorously to mix the contents.

5. Add 5 mL of 2% pancreatin solution and 2 mL of 5% bile salt solution to the third test tube. Cork the tube with a cork stopper and shake vigorously to mix the contents.

6. Using a pH wand or pH meter (as demonstrated by your instructor), measure the initial pH of the contents of each tube and record that value on the report sheet. Clean the pH probe between measurements by rinsing with detergent solution and then with distilled water.

7. Place the tubes in a 37°C water bath and allow them to sit undisturbed. Make sure the tubes are supported and will not fall over in the water bath.

8. Repeat the pH measurements at 20, 40, and 60 minutes. Record the pH values on the report sheet.

9. Carefully observe the contents of each tube after each pH measurement and record your observations on the report sheet.

10. Dispose of all solutions, test strips, and egg waste from today's experiment as directed by your instructor.

Name _____

Date _____ Lab Section _____

Pre-Lab Questions | 15

1. Pull the center out of a slice of fresh white bread. Roll it into a loose ball and place it in your mouth. Chew the ball a couple of times, but do not swallow it. Allow it to sit in your mouth for a minute or so. Do you notice any change in the taste? If so, describe the change and explain what produced the taste.

2. What type of chemical reactions do digestive enzymes catalyze?

3. Explain the purpose of conducting a control experiment in a scientific investigation.

4. What is the function of bile in the digestion of fats?

Name _____

Date _____ Lab Section _____

How Is Food Broken Down During Digestion?

Part 1. Digestion of Proteins

Observations of Egg White

Time/Tube	H_2O + HCl	Pepsin + HCl	Pepsin + H_2O
Initial			
15 min			
30 min			
45 min			
60 min			

Part 2. Digestion of Starch

Observations of Glucose and Iodine Tests

Time/Tube	Starch + Amylase		Starch + Water	
	Glucose test	Iodine test	Glucose test	Iodine test
Initial				
5 min				
10 min				
15 min				

Part 3. Digestion of Fats

Measurement of pH and Observations of Cream Digestion

Time/Tube	Bile salts	Pancreatin	Pancreatin + Bile
Initial	pH_____	pH_____	pH_____
20 min	pH_____	pH_____	pH_____
40 min	pH_____	pH_____	pH_____
60 min	pH_____	pH_____	pH_____

QUESTIONS

1. Consider the reactions in Part 1:

   a. In which tube did you notice the most significant change in the appearance of the egg white (indicating digestion of the egg white protein)?

   b. Based on your results, did pepsin digest the egg white better in combination with H_2O or with HCl?

   c. What does this indicate about the optimum conditions for pepsin (refer to section 10.7 of your text) and its location along the digestive tract?

2. Consider the reactions in Part 2:

    a. Did starch break down in the water solution? Explain how you know.

    b. Was starch digested in the tube containing amylase? Explain how you know.

    c. Did you detect glucose in either of the tubes? What does this indicate about the products of the starch digestion?

3. Consider the reactions in Part 3:

    a. In which tube did the pH change most significantly?

    b. What product of fat digestion causes the observed change in pH?

    c. Which tube's contents showed the most significant visual change?

    d. Did the bile salts facilitate fat digestion? Justify your answer.

Appendix

SIGNIFICANT FIGURES AND MEASUREMENT

Chances are that in a previous chemistry course or in this class, you encountered the concept of significant figures. You may have learned how to distinguish whether a certain digit within a given number is significant. Perhaps you learned how to determine the correct number of significant figures in a number you obtained by adding, subtracting, multiplying, or dividing other numbers. What you may not have learned is why significant figures are important.

Significant figures are important because of measurement. Simple as that. Measuring any quantity such as time, length, mass, or volume results in a number with a particular number of significant figures due to the precision of the instrument used to measure the quantity. We say that the significant digits in a number are those we know with certainty and the digit that is estimated. Let's illustrate with an example. You will need a U.S. 25 cent piece (a quarter) for this example.

Using the ruler below, measure the diameter of a quarter. Record your measurement in the blank provided.

The ruler you used here had graduations only for the centimeter marks. Therefore, you could determine that the diameter of the quarter was more than 2 cm but less than 3 cm. You know that the diameter is at least 2 cm; so this first digit you know with certainty, and it is significant. Then you should have estimated the next digit of your measurement. Perhaps you estimated that the edge of the quarter was halfway between the 2 and 3 cm marks and wrote your measurement as 2.5 cm. With no graduations between the 2 and 3 cm marks, you cannot know the exact position of the edge of the quarter between these marks—the ruler does not measure with that level of precision. However, you can estimate the approximate position of the edge of the quarter, which means that the digit you recorded is significant. Your recorded measurement of 2.5 (or 2.4 or 2.6) cm has two significant figures. Any measurement made with this ruler can have no more than two significant figures.

Now measure the diameter of the quarter using the ruler below. Record your measurement in the blank provided.

This ruler has the graduations for the centimeters as above, but also includes the graduations of millimeters. In other words, this ruler has greater precision than the one above. With this ruler, you can see that the diameter of the quarter was greater than 2.4 cm but less than 2.5 cm. Depending on the age of the quarter and the amount of wear, you probably measured a diameter that was slightly more than 2.4 cm. But how much more? In your measurement, this is the estimated digit and is significant. So whether you measured 2.41 cm or 2.43 cm, your measurement has three significant figures—the first two digits, 2 and 4, that you know with certainty along with the digit you estimated.

As this example illustrates, significant figures are not some evil entity created to torment chemistry students; instead, they arise from the precision of the instrument used to take a measurement. The more precise an instrument, the greater the number of significant figures possible in the measurement. Keep in mind that in any measurement, the significant digits are those digits that are known with certainty along with the estimated digit of the measurement.

For more information on counting the number of significant figures in measurements, please refer to Section 1.3 of *General, Organic, and Biological Chemistry, 2e.* The table below summarizes the rules for counting significant figures in measurements.

TABLE A.1 Counting Significant Figures in Measurements

Rule	Measurement	Number of Significant Figures
1. A digit is significant if it is		
a. not a zero	41 g	2
	15.3 m	3
b. a zero between nonzero digits	101 L	3
	6.071 kg	4
c. a zero at the end of a number with a	20. g	2
decimal point	9.800°C	4
2. A zero is not significant if it is		
a. at the beginning of a number with a	0.03 L	1
decimal point	0.00024 g	2
b. in a large number without a	12,000 km	2
decimal point	3,450,000 m	3

SIGNIFICANT FIGURES AND CALCULATIONS

Just as a measurement can be no more precise than the instrument used to take the measurement, the answer obtained from a mathematical equation cannot have more significant figures than the numbers used to obtain the answer. For example, if you measure two lengths—one of 12 cm and one of 3.13 cm—you can see that the first measurement was taken with a less precise instrument. When the two numbers are added together, the addition cannot increase the precision. So while the sum is 15.13 cm, the answer to the correct number of significant figures is 15 cm. The less precise of the two measurements determines the number of significant figures in the answer. For addition and subtraction, consider the places to the right of the decimal (if any) to determine the precision of a measurement.

> 12   cm   (no places after the decimal)—this is the less precise number
> +3.13 cm   (two places after the decimal)
> ———
> 15.13 cm   (a calculator gives us this number)—we must use same number of decimal places as the less precise number, so we round the number to 15 cm

When multiplying or dividing measurements, the concept is the same (the answer can be no more precise than the least precise number), but the application of the rule appears slightly different. Instead of looking at the number of places to the right of the decimal point (as we did above), we consider the number of significant figures in each number. For example, if you were to convert a measurement of 10 in. to centimeters, you would multiply 10 in. by 2.54 cm/in. to get 254 cm. While this may seem logical, recall that we only know the first measurement to one significant figure. It is actually 10±1 inch. So we appear to have gone from knowing that the measurement is somewhere between 9 and 11 in. to knowing that it is *exactly* 254 cm. The mathematical operation cannot increase the precision of the measurement.

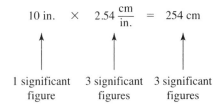

10 in.   ×   2.54 $\frac{cm}{in.}$ =   254 cm

1 significant   3 significant   3 significant
figure   figures   figures

Because the first number has only one significant figure, the answer can have only one significant figure and is rounded and reported as 300 cm. The answer must have the same number of significant figures as the less precise number or the one with the least number of significant figures in the equation.

RULES FOR ROUNDING NUMBERS

For our purposes, we will use the following convention for rounding numbers. Begin by determining the final digit to be retained; then look at the number to the right of it (the first digit to be dropped):

If that number is 4 or less, simply replace it and all remaining digits to the left of a decimal point with zeros.

If that number is 5 or greater, increase by 1 the final digit to be retained and replace the first digit to be dropped and all remaining digits to the left of a decimal point with zeros.

Digits to the right of a decimal point that are not to be retained are simply dropped.

During multiple-step calculations, keep all of the digits until the end of the calculation. Rounding after each step of the calculation introduces rounding errors and produces incorrect answers.

In the calculation above, 254 was to be rounded to 1 significant figure. Meaning only the first digit (in the place occupied by the "2") was to be retained. Since the first digit to be dropped was a "5," the digit to be retained is increased by 1, and zeros were substituted as placeholders for the remaining digits (which were both to the left of the decimal).

The following example illustrates rounding a number with non-retained digits to the right of the decimal.

    23.781 cm

    +4.01 cm

    27.791 cm   (answer can only have 2 places to the right of the decimal)

The digit where the "9" resides is to be retained. Since the number to its right is 1, the "9" is not changed, and the "1" is simply dropped; it is not replaced with a zero. The correctly rounded answer to the correct number of significant figures is 27.79.

NOTES

NOTES

NOTES

NOTES